International Environmental Risk Management

ISO 14000 and the Systems Approach

John Voorhees
Robert A. Woellner

LEWIS PUBLISHERS

Boca Raton Boston New York Washington London

Library of Congress Cataloging-in-Publication Data

Voorhees, John.
    International environmental risk management: ISO 14000 and the
systems approach / by John Voorhees and Robert A. Woellner.
        p.   cm.
Includes bibliographical references and index.
ISBN 1-56670-291-7 (alk. paper)
    1. Environmental risk assessment--United States.   2. Environmental
management--Standards--United States. 3. ISO 14000 Series
Standards.   I. Woellner, Robert A.   II. Title.
GE145.V66   1997
363.7′05--dc21

                                                                            97-29139
                                                                                    CIP

No claim to original U.S. Government works
International Standard Book Number 1-56670-291-7
Printed in the United States of America   1 2 3 4 5 6 7 8 9 0
Printed on acid-free paper

Preface

The Dawn of the
Environmental Age of Reason

"Think global, act local" is a catchy, insightful phrase sometimes affixed to automobile bumpers in the U.S. that raises the public's consciousness of their role in protecting the environment. In the business world, global environmental standards like ISO 14000, developed by the International Organization for Standardization (ISO), are the functional equivalent of such slogans as they address environmental concerns on a worldwide basis. These standards are now being employed by national and multinational businesses to manage and control environmental risks and liabilities, thereby producing better performance and reducing adverse impacts on the ecosystem. Long-lasting solutions to global environmental problems need the attention and dedication of the people causing impacts, both positive and negative, on our environment. This book shows how environmental problems are resolved by environmental management systems that use ISO 14000 standards coupled with other techniques to reduce environmental impacts and simultaneously create revenue growth.

Governments can be helpful in assisting businesses to achieve environmental excellence. Twenty years ago companies were rebelling against government regulations designed to force businesses to make changes to improve worker safety and environmental performance. In 1980 the U.S. Supreme Court rejected claims by the textile industry that Occupational Safety and Health Administration (OSHA) regulations would destroy its business. The Court did not dispute the adverse financial impacts, but rather allowed the regulations preventing workers from inhaling cotton dust on factory floors. The textile industry responded by creating new high-speed equipment that not only reduced dust and downtime by cleaning dust from the machinery, but also manufactured fabric at many times the rate of the old, unhealthy process (Lavelle, 1995). The textile industry did all of this at one-third the estimated cost and bolstered the economic viability of a faltering industry. The lessons learned in the textile industry with OSHA regulations are now being transferred to other industries to solve pollution problems by using new management systems designed to identify and control existing risks, significantly reduce future risks and liabilities, and most importantly, minimize impacts.

While governmental regulations can be helpful to businesses, governmental intervention can also retard environmental progress and waste resources that could be used to prevent rather than control pollution. In the U.S. during the 1980s and 1990s, tough environmental regulations were routinely fought by industry and frequently produced disincentives for businesses to minimize their environmental impacts. Former Environmental Protection Agency Administrator William K. Reilly has criticized the U.S. regulatory approach because it was unsuccessful in solving the greatest environmental problems: "We've had Love Canal, Valley of the Drums, and the Exxon Valdez. With virtually every case of a new environmental crisis, there is a new legislative priority and a new budget allocation. That has created a mix of programs that doesn't respect the biggest risks to health and ecology" (Schneider, 1993a). By reacting to environmental crises without considering the full range of consequences of regulatory policy, we have made costly mistakes. The mistakes need to be rectified quickly and scarce resources must be reallocated to find permanent solutions to environmental problems throughout the world.

In 1982 government officials discovered high concentrations of dioxin in dirt roads in Times Beach, MO. The federal government caused a near panic by evacuating and permanently relocating 2240 residents at a cost of $37 million. A decade later the federal official who ordered the evacuation admitted that he had made a mistake because he had overestimated the danger presented by dioxin (Schneider, 1993a). In another example in 1985, Congress passed a law requiring local and state governments to remove asbestos from public buildings at a cost of $15-20 billion. In 1990 the EPA admitted that the removal of asbestos frequently increased rather than decreased the asbestos exposure for building occupants (Schneider, 1993a). These are just two examples of wasted financial resources that could have been used to permanently improve our environmental quality of life.

Governments can be expected to make mistakes, particularly involving the environment where we still have limited knowledge as to the long-term effects of chemicals. Businesses, however, can not afford to commit significant errors when it comes to environmental protection. By requiring perfection from businesses in complying with rigid pollution control standards, we have set ourselves up for failure. The message sent to business is that it is more important to be in current compliance than to search for ways to dramatically reduce environmental impacts. Business leaders need to work with government to improve their performance and, at the same time, identify and improve regulations that waste resources so that funds can be reallocated to solve real environmental problems. In Yorktown, VA the Amoco Oil Corporation was required to spend $31 million to rebuild the refinery's wastewater treatment plant to prevent benzene, a toxic chemical, from evaporating into the air. A study jointly funded by Amoco and the EPA found that federal regulations did not cover the major source of benzene emissions at the marine terminal of the refinery which could have been abated for $6 million (Schneider, 1993b). By working together, as opposed to against each other, Amoco and the EPA solved an environmental problem that regulations had missed.

The *Ocean Dumping Act* is an example of well-intentioned pollution control legislation that does not address the real environmental problem. In the blistering summer of 1988, health officials closed beaches in New York and New Jersey after discovering bloody syringes, infectious hospital wastes, and dead dolphins scattered on beaches. Horrified citizens clasping pictures of disgusting debris descended on Congress, which then acted swiftly. In the Fall of 1988, Congress unanimously passed the *Ocean Dumping Act* which prohibited New York City from dropping its processed waste into the sea. However, the contaminated beaches were not caused by sea dumping but rather by the ancient combined sewer overflows in the metropolitan area, causing raw sewage and other noxious wastes to be flushed into the sewers after heavy downpours. Congress' solution required New York to spend vast sums of money landfilling the processed waste, but that did not solve the problem. While one politician gleefully declared "This is a turning point in human history," in actuality it was like the Amoco refinery, just another example of government's misidentification regarding the source of the environmental problem.

In 1990 the EPA's Science Advisory Board concluded that environmental laws "are more reflective of public perceptions of risks than of scientific understanding of risk" (Schneider, 1993a). Cataclysmic events in the late 1970s and 1980s created extraordinary public awareness of environmental problems and a demand for government action. Incidents like the Bhopal tragedy, the Exxon Valdez oil spill, and the discovery of toxic waste dumps produced tough legislative restrictions on new development, existing business operations, and the disposal of all wastes. These events are important because they captured the media's attention and molded the public's viewpoint that environmental risks are unacceptable in the modern age.

The government's pollution control strategies have also created many unwanted and adverse collateral consequences. Major environmental legislation in the U.S. has created a huge industry for lawyers and consultants to spend corporate resources squabbling with the government over past disposal practices and conformance to strict regulations. "Policymakers, business leaders, and environmentalists have focused on the static cost impacts of environmental regulations and have ignored the more important offsetting productivity benefits from innovation. The whole process has spawned an industry of litigators and consultants that drain resources away from real solutions" (Porter and van der Linde, 1993a).

In the past decade, governments have concluded that they do not have the resources, the personnel, or the technical ability to solve all of the global environmental problems. Well-publicized efforts by the U.S. government to put polluters in jail have stalled because the EPA does not have the financial resources or personnel to police the nation's criminal environmental laws. The European Environmental Agency has a minuscule $18 million budget, half of which is spent on outside consultants who monitor environmental conditions in Europe. The European Union (EU) created this agency with no sweeping investigative powers and without the responsibility to inquire into its members' internal affairs. Contrast this with the large multinational companies that are currently spending billions on pollution control measures each year (*The Economist*, 1991). In the U.S. $150 billion is spent annually by businesses trying to comply with environmental regulations. (Mcinerney and White, 1995) Total worldwide expenditures for environmentally related products and services will soon top $500 billion, exceeding the car market and reaching half the size of the information technology market. By some measures, the environmental industry is growing faster than biotechnology, telecommunications, or computer software (Mcinerney and White, 1995).

It is corporations, not governments, that are the only organizations with the financial and human resources, the technology, the global reach, and the motivation to achieve sustainability in future generations (Hart, 1997). These funds can be used to control pollution, but more importantly they should be used to eliminate the causes of pollution, reengineer production processes, and cut costs. Businesses must learn how to turn pollution control strategies into environmental opportunities that become major sources of revenue growth. "Focusing on sustainability requires putting business strategies to a new test. Taking the entire planet as the context in which they do business, companies must ask whether they are part of the solution to social and environmental problems or part of the problem" (Hart, 1997). Swift action by businesses is needed now to avoid the same mistakes of the past. There is already great competition for the scarce resources that will be allocated to clean up the past problems, and businesses cannot afford to make mistakes in the future that can add to the existing contamination on the earth.

The disposal of toxic chemicals is only a small part of a much greater problem that is harder to portray on a television screen, but far more likely to have an enduring, negative impact on the environment and business — resulting in greater challenges in the 21st Century. Anthony D. Cortese, the President of Second Nature, a nonprofit organization established to improve environmental education, argues that rapid population growth presents a fundamental danger to our environment (Cortese, 1994). In 1830 the population reached 1 billion, by 1950 it grew to 2.5 billion, and then doubled by 1988. As the world's population increases by 90 million people each year, 25 billion tons of topsoil are lost due to deforestation: the equivalent of all the wheatfields in Australia (Anderson, 1995). Of the 5.8 billion people that inhabit the world, approximately 1.5 billion live in conditions of abject poverty (Magretta, 1997). One out of every three people lives in a city. By 2030 when the population doubles again, two out of three people will live in a city. There will be over 30 megacities with populations of 8 million people or more and 500 cities with populations exceeding 1 million (Hart, 1997). Unless there are extraordinary measures taken immediately on how we deal with our environment, "the kind of world implied by those numbers is unthinkable" (Magretta, 1997).

As the population has increased, so has industrial production that causes pollution and depletion of natural resources as business responds to marketplace needs. World industrialization has increased 100 times in the past 100 years (Clark, 1989). With increased production comes more industrial pollution and wastes piled on top of the remnants of unregulated disposal of past refuse. Consumers are using and discarding products in greater numbers than ever before.

Nations throughout the world are being forced to change how they protect the environment based upon the growing population's escalating environmental use.

Acid rain is a major problem in cities and countries where coal combustion is unregulated. There is no country in the world that has solved its air and water pollution problems. All have stockpiles of toxic chemicals leaching in waste dumps. Equally threatening is the consumer's use of products, such as automobiles, that contribute to immense pollution throughout the world. By 2010, according to the World Bank, there will be more than one billion automobiles in the world, mostly concentrated in cities where they will double current levels of energy use, smog precursors, and emission of greenhouse gases (Hart, 1997). The growing dependency on cars in U.S. cities like Los Angeles and Asian cities like Bangkok and Kuala Lumpur has created virtual gridlocks in the absence of adequate transportation planning and infrastructure improvements. One economic expert estimates that traffic congestion costs an extra $200 billion a year in wasted fuel as automobiles and trucks idle on the highway, lose time, have accidents, and create other related wastes (Hawken and McDonough, 1994). Our increasing use of the automobile is just one of many factors that needs to be addressed globally if we can hope to reduce our environmental impacts and create a healthier, more sustainable environment.

Innovative educational efforts, begun by Anthony Cortese in the U.S. and Paul Hawken in Sweden, are demonstrating how people can be taught to become better caretakers of the environment. Cortese argues that we must change the way we educate people on environmental issues to create more responsible stewards of the environment for generations to come. Cortese and Hawken are urging schools to develop environmental education-awareness curricula to be added to all subjects taught in secondary and higher education. By making environmentalism second nature to future business and government leaders, we can begin to change how products are produced and used, how wastes are created and discarded, and how people can voluntarily minimize their individual environmental impacts. Cortese believes that the root cause of our neglect of the environment for many years was that as a nation and as a world, we believed our individual and collective impacts would result in only imperceptible environmental changes. This anthropocentric world view caused people to believe that they were the most important biological species and thus should have dominion over all of nature. People assumed that natural resources were free and inexhaustible; wastes could be assimilated in the environment; and societies and continents could adapt to any adverse changes (Cortese, 1994). Years of production of nuclear weapons in during the Cold War increased the legacy of environmental neglect and degradation. In the U.S. the total bill just to clean up all the hazardous, chemical, and radioactive waste sites is somewhere between $700 billion to $1 trillion.

Political and social pressures are forcing businesses to pay extraordinary costs of doing business and using the environment. The definition of business and development costs is rapidly expanding to include new costs that have never been borne by business. For example, many people believe cigarette manufacturers should charge $3.63 for a pack of cigarettes, which is, according to one study conducted by the University of California at San Francisco the combined costs of the cigarette and the adverse health consequences from smoking. It is just a matter of time before the cigarette manufacturers who have settled massive products liability cases in the U.S. in 1997, will be required to create a fund for the health costs of smoking. Governments throughout the world are considering green taxes on carbon dioxide and other forms of pollution. Singapore already charges people to drive in its metropolitan center and has imposed quota premiums for car buyers. The right to buy a car is auctioned off to the public, and can cost up to $20,000 for a luxury car (Mcinerney and White, 1995).

Industrial and commercial expansion, and even residential growth, are being taxed at unprecedented rates and regulated by more government authorities than ever before. Citizens and environmental activist groups are erecting barriers to development and forcing businesses to become more responsive to their environmental and quality-of-life agendas. Conflicts are becoming less civil between assetholders, who have property interests and seek freedom from regulatory restraints, and stakeholders, who have use and enjoyment interests and urge greater access and less environmental degradation. Pollution knows no geographic, cultural, social, or political boundaries, and hence the line between assetholders and stakeholders is not precise. Under the modern theory of environmental dispute resolution, every interested party wants a seat at the bargaining table with an equal vote and a meaningful opportunity to be heard. While global issues are becoming more pronounced (e.g., global warming, population, and deforestation — to name a few) there are many competing and urgent local issues that are creating tensions and disagreements on how we should allocate scarce public funds to contribute to environmental solutions. The discourse is becoming increasingly volatile because the public is blaming governments and businesses equally for allowing adverse impacts to the environment. People are more intolerant of any form of pollution, believing that governments and businesses owe them the duty to make their lives free of risks posed by industrial development. Public censure, amplified by worldwide media channels, is becoming perhaps the largest operations problem faced by businesses today.

Business leaders know that pollution is a significant cost of daily business. When any type of waste is discarded into the environment, "it is a sign that resources have been used incompletely, inefficiently, or ineffectively" (Porter and van der Linde, 1995a). Pollution often reveals defects in product designs, manufacturing processes, inattentive management practices, and unresponsiveness to environmental issues. It is a red flag for poor quality (Mcinerney and White, 1995). Passing on the costs of pollution to the customer is a bad management practice that will be discovered and exploited by various interest groups. This inevitability will lead to loss of market share and public censure.

Environmental risks are inherent in business operations, and can become oppressive. Few businesses in the world are immune from the threat of massive legal and financial liability for environmental claims, coupled with the loss of corporate reputation. Whether true or false, claims can create disastrous consequences. Two environmental activists in London recently claimed in a leaflet that the McDonald's hamburger chain destroys rain forests, flouts environmental concerns with its packaging, sells dangerously unhealthy food, seduces children into unhealthy eating habits, exploits staff (particularly blacks and women), and is responsible for torturing animals (Montalbano, 1997). McDonald's responded with a libel suit that lasted 314 days and now holds the record as the longest trial in British history. McDonald's spent $16 million to win the case and recovered a paltry $98,000 in damages. The company's reputation, however, has suffered with the widespread attention the libel suit has received. The case was featured on a two-night, three-hour British television reconstruction of the case called "McLibel!" and a book titled "*McLibel: Burger Culture on Trial*"; and a World Wide Web site on the trial has been browsed by 9 million people (Montalbano, 1997). There are many other examples of companies whose reputations have been damaged because of actions taken by environmental activists to publicize negative environmental impacts.

The myriad of complex environmental laws and regulations now require responsible corporate officers to monitor operations more closely. Failure to discharge these new obligations can lead to disastrous financial consequences. Investor scrutiny directed toward corporate management regarding environmental matters has never been more intense.

In the U.S., following the Caremark International Inc.'s guilty plea in a criminal case which cost the company $250 million in related criminal and civil fines, Caremark shareholders sued the company's Board of Directors. They alleged that the Board bore personal responsibility for inadequately supervising the conduct of lower-level employees. The Board, however, was found to have properly discharged its duty because it had issued compliance guides and manuals that were reviewed and revised annually by outside advisors to the company; developed standard forms which assured compliance with applicable regulations; appointed a senior official to monitor and assure compliance; adopted an internal review process to assure that supervisors implemented compliance policies and procedures; evaluated compliance programs by an independent third-party auditor; issued a corporate code of conduct and ethics manual for all employees; and established a toll-free, confidential hotline for use in reporting infractions (Rothenberg, Smith, and Feitshans, 1997).

The Caremark decision has been widely publicized because of the implications to corporate executives and managers who must not only understand how environmental issues affect the operation of their businesses, but who also must correctly monitor them before damage occurs. Failure to exercise due diligence in taking steps to prevent and detect noncompliance can mean enhanced penalties pursuant to the U.S. Sentencing Guidelines for Organizations, but it is far more likely that it will result in a loss of market share, civil liability, and, perhaps, business failure. Preventing violations has become an important consideration in developing and implementing an environmental risk management and risk reduction program. Smart businesses seek to avoid these pitfalls by creating innovative solutions to pollution problems, and potentially increasing their market share in the process.

Two environmental authors concluded, after a comprehensive environmental impact study of ten global businesses, that "[c]ompanies that pollute the least have the highest quality products and services" (Mcinerney and White, 1995). Companies can expand their market share and lower costs by drawing the connection between waste, margins, and quality (Mcinerney and White, 1995). Systems that were developed for pollution control need to be shifted to pollution prevention. "Pollution control means cleaning up waste after it has been created. Pollution prevention focuses on minimizing or eliminating waste before it is created. Pollution prevention strategies depend on continuous improvement efforts to reduce waste and energy use" (Hart, 1997).

Environmental regulations have historically overlooked source reduction and focused almost exclusively on pollution control. Businesses responded by creating technologies that focus on achieving stringent standards without eliminating the causes of pollution. Recycling is one means that has been used to address the problem after it occurs. Recycling can be an expensive, labor-intensive way of dealing with waste that does not address the real problem of eliminating the source of the waste. Recycling is an important first step because it makes people more aware of the scope of our environmental problems and allows them to participate directly and meaningfully in a community-wide effort to mitigate waste. It is, however, a poor substitute for the elimination of waste in the production process.

Companies need to use design for environment, as a new tool for creating products that are easier to recover, reuse, and recycle, if necessary. Design for environment examines, during the design phase, all of the effects that a product could have on the environment. This includes the natural resources and energy used to create the product, as well as the packaging used in shipping the product to the consumer. This life-cycle product analysis begins and ends outside the boundaries of a company's operations, and examines how customers use and dispose of the products (Hart, 1997). The same type of analysis can be conducted for service businesses to review their internal operations and create a comprehensive plan to minimize environmental impacts.

We are now emerging into a new era of environmental reason where governments are learning how to create incentives for companies to innovate and improve product performance, quality, and management. Innovation can reduce the use of power, natural resources, and production materials. Businesses that take a hard look at their internal operations will notice opportunities for the development of effective environmental management systems. Those systems not only reduce liabilities and risks, because less pollution is produced, but they also increase profits because all resources are used more efficiently and effectively, making informed consumers more likely to buy the product or service even at higher per unit prices.

Professors Porter and van der Linde have argued persuasively that businesses now need to focus on outcomes, not technologies, by using systems approaches. Businesses must convince governments to enact environmental regulations that promote real innovation — the kind that business can recognize and implement rather than litigate in the court systems. Regulations should encourage upstream solutions to environmental problems and employ market incentives, drawing attention to resource and production inefficiencies. When the regulatory process becomes more stable and predictable, it can provide more opportunities and incentives for business to go beyond compliance and create revenue growth. Porter and van der Linde urge that industry must participate in setting standards from the beginning by developing strong technical capabilities among regulators and minimizing the time and resources consumed in the regulatory process.

New regulations alone are insufficient to drive business people to make correct decisions on how to manage and ultimately reduce their environmental risks and liabilities. Management systems must also be employed to show businesses how to take advantage of all the opportunities to redesign internal operations to limit environmental impacts. Cortese advocates gentle persuasion, not preaching, to change the way we do business to make meaningful environmental progress (Cortese, 1994). He contends that the three most important ways to change corporate behavior on environmental issues are (1) make the business leaders believe they will not fail at the new venture; (2) show them how other companies are doing the same thing; and (3) demonstrate that the new behavior will result in financial gain.

A systems approach using global ISO 14000 standards coupled with risk reduction techniques can assist in changing corporate behavior. Sustainability can be achieved if we make continual improvements of our processes and services and eliminate environmentally degrading sources of pollution. The systems approach can create sustainability and, at the same time, allow businesses to identify and limit their impacts, prevent pollution, restore natural resources, and minimize environmental risks and liabilities. Managing and reducing risks are part of a global ideological shift that is changing the way that businesses have operated and is creating opportunities for future generations to live free of past encumbrances.

The arena of corporate risk management has traditionally been characterized by two types of efforts: those performed to control existing risks and those performed to reduce future risks. Managing environmental risks effectively requires a comprehensive systems approach that combines both skills. The ISO 14000 series of standards can serve businesses as a model for an integrated series of management systems that identify, control, and monitor environmental risks. Additional risk management strategies include the use of information technology, risk-based approaches to internal and external environmental communication, insurance and risk-transfer methodologies and legal risk reduction strategies.

Reducing future risks has as its central focus the identification of the causes of potential regulatory, public, and employee conflicts coupled with the elimination of adverse environmental impacts. ISO 14000 environmental management systems need to be augmented with auditing, litigation avoidance, collaborative decision making, alternative dispute resolution, involvement with proactive EPA environmental programs, and other innovative risk reduction techniques. All of these areas need to be considered when creating or enhancing a comprehensive environmental risk management system.

The systems approach takes these risk managment and reduction strategies and techniques and creates opportunities for further process innovations and solutions to environmental problems which will make the world a healthier, safer, and more prosperous place in which to live. It employs methods to avoid repeating past mistakes while at the same time encourages future revenue growth. It provides strategies for corporate managers to follow when developing environmental risk management systems together with specific examples of techniques to eliminate risk. It contains guidelines for implementing environmental management systems and using daily experience with the system to enhance its effectiveness.

When environmental risk management systems are effectively implemented, they reduce pollution, minimize regulatory problems such as potential civil and criminal environmental liability, and maximize the public's and consumers' confidence in public safety and the businesses' efforts to preserve and protect the environment. The *systems approach* serves as a written guide for management and employees to conduct business operations in a safe, environmentally responsible manner. Most importantly, it demonstrates to the public that businesses are consciously and proactively managing their environmental concerns and are genuinely committed to safeguarding the environment.

ABOUT THE AUTHORS

John Voorhees is an environmental lawyer who resides in Boulder, CO. He directs the litigation department and the environmental law practice group of the Denver law firm of Isaacson, Rosenbaum, Woods & Levy, P.C. He is Chief Legal Officer of QUEST Management International, LLC, a firm with offices in Denver and London that develops ISO 14000 environmental management systems for businesses. He is an adjunct member of the faculty at the University of Denver College of Law and the author of numerous articles on environmental risk management and reduction strategies. He is a well-known lecturer in the U.S., Europe, and Asia on global environmental issues. He is a member of the National Commission for Corporate Compliance of the National Center for Preventive Law. He graduated from Emerson College with a B.A. degree in English and received his law degree from the Columbus School of Law of the Catholic University of America.

Robert A. Woellner is the President of QUEST Management International, LLC and Senior Vice President of Cherokee Environmental Risk Management. Mr. Woellner is an ISO 14000 Lead Assessor and a member of the U.S. Technical Advisory Group to ISO/TC 207. He is actively involved in assisting many businesses with the development and implementation of environmental management systems. Mr. Woellner has significant experience in environmental risk assessment and management, environmental insurance, analytical services, and the development and implementation of environmental management systems. He serves on several boards, teaches college classes, and has published many articles and contributed to several books regarding environmental risk management. He holds a B.S. in geology from Middlebury College and an M.S. in marine geology and geophysics from the University of Miami — Rosenteil School of Marine and Atmospheric Science.

The opinions expressed in this book are not necessarily the views of clients of Isaacson, Rosenbaum, Woods & Levy, P.C., or QUEST Management International, LLC.

ACKNOWLEDGMENTS

We deeply appreciate the efforts of our friends and colleagues who contributed immensely to the book. Special thanks to Edward A. Dauer, Trustee of the National Center for Preventive Law and Dean Emeritus, University of Denver College of Law, for his encouragement and insights regarding compliance systems and preventive law. David A. Fazzone of McDermott, Will and Emory, and Richard M. Burns, Thomas J. McBride, and J. David Mannion of Eastern Utilities Associates have made many helpful suggestions throughout the project. Dean Jeffrey Telego and Tacy Telego provided us with many opportunities to make presentations at their environmental conferences, which allowed us to develop new concepts amongst our peers. Don Burklew, Frank Faris, and James "Skip" Spensley gave us encouragement and advice on a range of subjects contained in the book. We are particularly thankful to John C. Kolojeski of Octagon Power Systems, Ridgway M. Hall, Jr. of Crowell & Moring, and Laura Belsten of the University of Denver for their help and encouragement regarding environmental compliance issues.

We would also like to thank Dennis Unites and Jerry Ackerman of GEI-Atlantic, who have worked with us in developing international environmental environmental management systems and brownfield redevelopment concepts using ISO 14000. Special thanks to Dr. Linda S. Spedding, Director of European Operations for QUEST Management International based in London, who is working with us developing international evironmental management systems. We believe Dr. Spedding represents the best of the new breed of international lawyers seeking solutions to global environmental problems. Roger Smith and Andrew Paine in London, and Mary-Rose Nguyen in the U.S., of Bureau Veritas Quality International have given us unqualified support on this project allowing us to become more familiar with the international registration process. We will surely miss Roger who passed away earlier this year. We appreciate the research assistance of Kirk Reiche and Gus Michaels, and also Kate and Howard Kilguss's insights regarding litigation avoidance, alternative dispute resolution and the redevelopment of brownfield sites. Thanks to A.J. Grant of Environmental Communications Associates for her help on internal and external corporate communications. We would like to acknowledge all of our friends at the Colorado Center for Environmental Management, including Molly Mayo and Shaun Egan, now of the Stapleton Development Corporation; the Denver Metro Chamber of Commerce Environment Committee, and the staff at the Chamber including David Ferrill, Elizabeth V. Woodward, and Tami D'Amico; and all the members of the E-net including Jamie Harrison, Jenifer Heath, Dan Himelspach, Tim Holeman, and Beth Smith for their helpful suggestions on environmental policies, alternative dispute resolution, and collaborative decision making. We would also like to thank our friend and colleague Anton Camarota of QUEST Management International for his significant contributions and input during this project; Jon Steeler of Isaacson, Rosenbaum, Woods & Levy, P.C. for his insightful and critical commentary; Sandy Blanch, Sherri Cordova, Linda Dwyer, Dorrie Emmerich, Sharon Lake, Patricia Murphy, Betty Rivas, and Juliet Wagner for their extensive work on production; and Michele Salazar for her outstanding efforts in review, production, and coordination of the entire project.

Finally, we would like to thank our wives Terri-Jo Woellner and Kim E. Voorhees for all their support.

For Our Parents

Mr. and Mrs. Theodore Voorhees

and

Dr. and Mrs. Richard C. Woellner

TABLE OF CONTENTS

PART ONE:

RISK MANAGEMENT STRATEGIES

Chapter 1

ISO 14000 AND RISK MANAGEMENT SYSTEMS

It is no coincidence that over 100 countries have joined together to create international standards for their industries and businesses to effectively manage their environmental impacts. The trend is towards universal solutions to environmental problems. Global environmental management standards are addressed in the first nine chapters, in which we show how to use the standards proactively to manage impacts. The next eight chapters are devoted to using additional legal strategies and voluntary initiatives to reduce environmental risks. The development of global standards is the appropriate phase to begin to see how responsible solutions are created to address worldwide environmental concerns.

In 1946 the International Organization for Standardization (ISO) was founded as a worldwide federation to promote the development of international manufacturing, trade, and communication standards, thereby facilitating the international exchange of goods and services (Hall, 1996). The ISO organization itself is a private sector, international standard body based in Geneva, Switzerland. ISO reviews input from government, industry, and other interested parties before it develops a standard. Understanding ISO and its mission and techniques points the way toward an appreciation of why the 9000 series for total quality management has been so popular, and why the 14000 series portends an even greater acceptance throughout the world.

ISO'S ROLE AND MISSION

ISO's mission to develop manufacturing, trade, and communications standards assisted in the rebuilding of Europe after the Second World War. As ISO grew rapidly, its purview became global and quickly involved nations outside of Europe. The members of ISO are the standards organizations from each nation, and current membership stands at more than 120 countries. The U.S. is a full voting member of ISO, officially represented by the American National Standards Institute (ANSI). ISO strives to systematically develop globally accepted standards that are voluntarily adopted by the international business community. Specific application methods and techniques are not specified, but are left to the discretion of industry and government experts from around the globe.

The mission of ISO has two parts:

- To promote the development of standardization and related activities in the world to facilitate the international exchange of goods and services,
- To develop cooperation in the sphere of intellectual, scientific, technological, and economic activity.

Initially, this mission focused on technical performance specifications for products and standardized test methods. Currently, more than 200 technical committees are dedicated to the continual development of these types of standards. In 1979, however, a number of worldwide market trends led to a change in focus for ISO. These included: the growth of industry throughout the world led by the post-WWII boom in the U.S.; the development of trade agreements and growth of international trade; and the proliferation of different quality standards throughout the world, including both product specifications and quality management systems.

The world markets were growing rapidly during the 1970s and 1980s, but were characterized by products and services that varied widely in their performance, characteristics, styling, materials, and interchangeability of parts. Since a key aspect of the ISO mission is to facilitate trade and remove trade barriers, ISO formed the Technical Committee (TC) 176 in 1979 to address these issues under the general topic of quality management. "The goal was to make it possible for purchasers in the international market place to ensure that products they bought were manufactured in accordance with known, verifiable, and accepted methods of controlling the manufacture and distribution of products" (Bell, 1995).

TC 176 was confronted by a bewildering array of quality standards for both product characteristics and quality management programs that had been institutionalized by various industrialized nations throughout the world. Such diverse standards constituted a set of technical barriers to trade. A wide variety of interpretations of the same standards, both throughout an industry and between industries, caused problems with products manufactured and traded worldwide. Industries generally agreed, however, that product specifications were not enough. Quality management had to address the processes used for purchasing, production, and servicing in order to assure attainment of uniform quality levels.

TC 176 set about the task of harmonizing the various quality management systems standards throughout the world, and issued the first *Total Quality Management Standards*, the ISO 9000 series, in 1987 (Hall, 1996). A second revision is currently underway. This series addresses the processes used by a business to ensure that it meets customer requirements for its products and services. The 9000 series does not define specific product performance levels or physical characteristics, but describes how to manage a business with a quality focus, thus attaining consistent results and providing confidence to the customer.

The 9000 series, as the forerunner of the 14000 series, has proved to be extremely popular throughout the world. Currently, more than 100,000 businesses worldwide have achieved registration, which consists of an independent third party (a registrar) declaring that a business' quality management systems conform to ISO 9000 requirements. In the U.S., Mexico, and Canada, there are more than 11,000 registered businesses, and that number is growing at a rapid pace (Bell, 1995). Approximately 1000 facilities per month are seeking certification (Hall, 1996). Some governments have even made the ISO 9000 series mandatory for businesses in their countries.

Worldwide, achievement of ISO 9000 registration has become a prerequisite for doing business in dozens of countries. Those businesses wishing to enter the European markets need to consider an investment in an ISO 9000 quality management system as an essential component of their business. ISO 9000 is a legal requirement in the medical devices market in the EU. It is also required in many other markets where product risk is a factor, such as high-pressure valves and public transportation. Competitive pressure is the primary reason for adopting ISO 9000, as thousands of firms are placing themselves on preferred supplier lists and demonstrating their global commitment to quality.

ISO 14000 SERIES OF STANDARDS FOR ENVIRONMENTAL MANAGEMENT SYSTEMS

Many of the trends that resulted in the emergence of the 9000 series also played a part in the development of the 14000 series. They include the growing international markets, the proliferation of environmental management standards and regulations in various countries, and the environmental management programs being adopted by businesses in response to complex environmental regulations.

Several issues related to the formation of the ISO 14000 series were different from those of ISO 9000. Any attempt to standardize environmental performance worldwide involves considerable social and political controversy. Implementing environmental management systems in countries with social democracies, such as Norway, has been relatively easy. In countries such as the U.S., the often antagonistic and litigious relationship between the government and the regulated community has caused environmental issues in the past to be approached by business people with extreme caution, if not fear (Begley, 1996).

Two trends in environmental management have been emerging as driving forces for ISO 14000 during the past two decades. In 1972 the United Nations held a conference on the environment in Stockholm, Sweden. A later conference was held in Rio de Janeiro, Brazil in 1992. The world community came together at each gathering to meld the views of diverse and sometimes opposing groups into a firm commitment to responsible environmental management and global sustainability. For the first time, the world established the environment as a priority in national and international affairs.

A seemingly opposing force had coalesced in the 1986 Uruguay round of the General Agreement on Trade and Tariffs (GATT). These negotiations resulted in a commitment to foster international trade. The Agreement on Technical Barriers to Trade section of the GATT does, however, encourage the use of international standards and conformity assessment systems in order to improve the efficiency of production and facilitate trade. This treaty requires that these international standards are not prepared, adopted, or applied with a view to or with the effect of creating unnecessary obstacles to international trade. Indeed, the technical regulations can not be more trade restrictive than necessary to fulfill a legitimate objective such as national security requirements; the prevention of deceptive practices; protection of human health or safety, animal or plant life or health, or the environment. This treaty establishes what could be perceived as an anti-environmental force by positioning international trade as a competing priority with environmental protection.

As a response to these emerging trends in environmental management, and cognizant of the resounding success and worldwide adoption of the ISO 9000 standards on quality management systems, in August 1991 ISO established a Strategic Advisory Group for the Environment (SAGE). This advisory group's purpose was to assess the need for environmental management standards and to recommend an overall strategic plan to develop these standards. ISO requested SAGE to consider the following issues:

- Promote a common approach to environmental management similar to total quality management standards (ISO 9000).
- Enhance businesses' abilities to attain and measure improvements in environmental performance.
- Utilize international standards to facilitate trade and remove trade barriers.

SAGE was specifically instructed not to consider environmental criteria, such as levels of pollutants, health assessments/risks, technology specifications, or product/process criteria. For over a year, SAGE studied the U.K.'s BS 7750 and other national environmental management standards as possible starting points for an ISO version. In 1993 SAGE recommended the formation of an ISO technical committee dedicated to the development of a uniform international Environmental Management Standard, as well as other standards on environmental management tools. ISO formed Technical Committee (TC) 207 to develop a series of environmental management system standards to accomplish the standardization in the field of environmental management tools and systems.

In June 1993 TC 207 met for the first time in Toronto, Canada where some 200 delegates representing approximately 30 countries agreed to complete a draft of the Environmental Management Standard and international auditing standards. SAGE was officially disbanded. Following an interim meeting on April 17-20, 1994 in Surfer's Paradise, Australia, on June 24-July 1, 1995 TC 207 met in Oslo, Norway, where 600 delegates representing over 50 countries agreed to elevate the environmental management standard and auditing standards to draft international standards with scheduled publication by the end of 1996. By July 1995 TC 207 had members from 63 countries.

Six technical subcommittees (SCs) and working groups (WGs) were created within TC 207 and are currently formulating standards in the following areas:

- **SC1: Environmental Management Systems, with the U.K. as the secretariat,** administered by the British Standards Institution
- **SC2: Environmental Auditing and Related Environmental Investigations,** with the Netherlands as the secretariat, administered by the Netherlands Normalisatie - Instituut
- **SC3: Environmental Labeling, with Australia as the secretariat,** administered by Standards Australia
- **SC4: Environmental Performance Evaluation**, with the U.S. as secretariat, administered by ANSI
- **SC5: Life-Cycle Assessment**, with France as the secretariat, administered by Association Francaise de Normalisation
- **SC6: Terms and Definitions,** with Norway as the secretariat, administered by Norges Standardiseringsforbund
- **WG1: Environmental Aspects in Product Standards,** with Germany as the secretariat, administered by the Deutsches Institut für Normung e.V.

On June 17-21, 1996 TC 207 delegates from 50 countries and 10 liaison organizations met in Rio de Janeiro, Brazil to complete the ISO 14000 drafting process. The main outcomes of this meeting were the following:

The 14001 and 14004 environmental management systems standards received final approval and were issued by ISO in September 1996. The 14001 environmental management systems specification was also approved.

The 14010, 14011, and 14012 environmental auditing standards were in the final stages of publication and had been overwhelmingly approved by the plenary. These standards were published in conjunction with the 14001 and 14004 standards in September 1996.

SC1 established a working group to gather information about the implementation of ISO 14000 in small- to mid-sized enterprises (SMEs). The working group will recommend the direction the TC should take towards developing the guidance standard, *14002 Environmental Management Systems — Guidelines on Special Considerations Affecting Small and Medium Size Enterprises.*

SC3 voted to begin work on Type III environmental labels, or what some call "environmental report cards." These labels are similar to the nutrition panels found on cereal boxes and are intended to provide the consumer with the necessary information when choosing a product based on its environmental impacts.

SC3 also reaffirmed the *14024 Environmental Labels and Declarations — Environmental Labeling Type I — Guiding Principles and Procedures* document as a committee draft, which places it in line for draft international standard status during 1997 or 1998. This standard could become the basis for mutual recognition among the 24 environmental labeling programs currently operating around the world.

SC4 decided to completely revise the existing environmental performance evaluation working draft and perform a line-by-line analysis of all comments received. The final result of this work was a fifth draft that reflected major advances in negotiations.

SC6 voted to elevate 14050, *Terms and Definitions,* to draft international standard status. This document should provide the user of the 14000 series of documents with a single source of terms employed in the practice of environmental management.

PRINCIPLES USED IN DEVELOPING INTERNATIONAL STANDARDS

To understand how the 14000 series is being developed and how these standards can be applied, ISO follows three basic principles in developing all standards: consensus, encouragement of full participation and voluntary adoption.

Consensus is a difficult task when one considers the complexity and controversy surrounding environmental issues worldwide. The development process for ISO standards, however, ensures that consensus is achieved through these basic steps:

Preparing a Justification. An appointed committee, usually part of a subcommittee, prepares a justification for any proposed standard, and formally submits a New Work Item proposal for a vote to the entire technical committee or the relevant subcommittee. If the majority of participating members vote in favor of the proposal and at least five members declare their commitment to actively support the project, development of the new standard proceeds.

Preparation of Working Documents. Groups of experts develop working drafts of the new standards, and refine them so they can advance to the next phase. It is in this phase where controversies are discussed, contentious issues are resolved at the most basic level, and a general consensus is reached among the experts.

Preparation of Committee Drafts. The working documents are formalized as Committee Drafts (CDs) and distributed for comments to the entire technical committee. Comments are reviewed and included, and more committee drafts produced as necessary until consensus is reached among the technical committee members and the document is ready to proceed to Draft International Standard (DIS) status.

Preparation of Draft International Standard. The DIS is circulated to all ISO member bodies for voting and comment within a period of six months. If a two-thirds majority of the participating members approve the standard and not more than one quarter of the members disapprove it, then the standard is moved forward to publication.

Consensus is achieved first by the technical experts, then by the technical committee, and finally by all the standards organizations and their experts throughout the world. Thus, consensus is built-in as a key aspect of the standards development process.

Participation in technical committees is open to all qualified and interested individuals. Industries that are likely to be affected by the standards are often involved with ISO technical committees as members of their national standards bodies. Adoption of these standards by countries and industries is voluntary, and based on market-driven needs. Thus, any businesses anywhere in the world can voluntarily choose to adopt these standards if it helps them to attain their vision or pursue their mission, based on the judgment of management. Businesses and their managers can be assured that the standards contain a high degree of technical integrity, resulting from the consensus of both industry experts and national standards bodies from more than 120 countries.

GOALS OF ISO 14000

The goal of ISO 14000 is to evolve a series of generic standards that provide business management with a structured mechanism to measure and manage environmental risks and impacts. Two major themes throughout the ISO 14000 standards emerge: "First is the desire for consistency in environmental management standards and practices. The idea is that wherever a business may be located, if the program complies with the standards, its excellence will be universally recognized" (Hall, 1996). The second important theme is that in designing an environmental management system each country or business "should be allowed the flexibility to consider its own processes, products and values, and to design for itself the most effective procedures to achieve and measure adherence to, or progress towards, the goals which are set forth in the program" (Hall, 1996).

Standards have been or are being developed for:
- Environmental management systems (ISO 14001-4)
- Environmental auditing (ISO 14010-14013)
- Internal reviews (ISO 14014)
- Environmental site assessments (ISO 14015)
- Evaluation of environmental performance (ISO 14031)
- Product-oriented standards such as environmental labeling, terms, and definitions for self-declaration environmental claims (ISO 14020-14024)
- Life-cycle assessment (ISO 14040-14043)

These standards are designed to help a business establish and meet its own policy goals through objectives and targets, organizational structures and accountability, management controls and review functions — all with top management oversight. The focus is on management rather than on performance standards. The centerpiece of this section, the 14001 standard for environmental management systems, provides a framework for assessing, managing, and reducing the liabilities associated with environmental aspects of operations. Through several key requirements of the 14001 standard, environmental management becomes a strategic, decision-making concern, allowing management to make more effective decisions for reducing risks. These same ISO 14000 standards, once they are developed and verified, can be a valuable management tool for businesses seeking to improve their environmental performance by reducing their environmental impacts and risks.

ISO 14000 is only one of many efforts being undertaken by businesses, environmental groups, investor organizations, governments and other entities to improve environmental performance of businesses. Nonetheless, ISO 14000 has a unique role because as an international standard, it will be given more credence and universal acceptance than its more parochial counterparts. For this reason, among others, it already fulfills an important function in the development of environmental management systems around the world. As the next chapter demonstrates, other entities also have had significant impacts on the development of environmental risk management and reduction systems and they must be carefully considered in conjunction with ISO 14000.

Chapter 2

OUTSIDE IMPACTS ON CORPORATE ENVIRONMENTAL RISK MANAGEMENT PERFORMANCE

While ISO was considering the development of a series of international standards for environmental management, pressures were mounting from worldwide environmental organizations like Greenpeace and national organizations like the Sierra Club, the Natural Resources Defense Council, the Environmental Defense Fund and others to make businesses more responsible for their environmental impacts. Professional organizations began developing their own guidelines on good environmental management practices. The Chemical Manufacturers Association created its Responsible Care® Program, adopted by all its member companies to improve the safety of handling and disposal of chemicals and otherwise to reduce environmental impacts. The Global Environmental Management Initiative (GEMI) developed and published its own set of guidelines based upon total quality environmental management principles, and addressed such topics as cost-effective pollution prevention, environmental reporting, environmental health and safety training, and benchmarking (Hall, 1996). The International Chamber of Commerce published its own set of environmental principles adopted by its members and other organizations to improve environmental decision making and reduce impacts. Even local business organizations began to get proactive. For example, the Denver Metro Chamber of Commerce developed its own environmental policy statement and a tool kit for its 3500 members to create their own environmental policies.

Other organizations decided to apply social, political, and investment pressures to change business practices. This chapter will highlight two public interest groups, CERES and CEP, that decided to force businesses to change their corporate cultures. The preliminary results of their campaigns are compared with the National Center for Preventive Law, an organization that assembled a group of experts to create voluntary corporate compliance principles and guidelines for businesses.

CERES

In 1989 an organization known as the Social Investment Forum brought together various environmental groups, institutional investors, government agencies and economists to make businesses adopt principles to govern their environmental performance. This group formed the Coalition for Environmentally Responsible Economies (CERES). The organizers believed that a collaboration between institutional investors and environmental groups could have a significant positive impact on the environment by pressuring businesses to reduce their environmental impacts. CERES has served as an impetus for getting businesses to adopt their own voluntary environmental management programs.

On September 7, 1989 CERES issued ten principles originally known as the *Valdez Principles*. These principles were named after the Exxon Valdez tanker that ran aground on March 24, 1989, spilling 11 million gallons of crude oil into the waters of Alaska's Prince William Sound. CERES modeled its Valdez Principles after the Sullivan Principles, written in 1977 by Reverend Leon H. Sullivan of Philadelphia in an effort to promote social justice and eliminate apartheid in South Africa. Within 5 years more than 140 multinational corporations, including Exxon, IBM, Citicorp, and Mobil, signed the *Sullivan Principles*. In contrast, the *CERES Principles*, which evolved from the *Valdez Principles*, in 8 years have been adopted by just 6 Fortune 500 companies. In 1996 CERES issued a report entitled *Reaching Critical Mass* in which it states that it is currently engaged in serious endorsement-related discussions with approximately 20 Fortune 500 firms and a dozen small- to mid-sized businesses and is expecting further endorsements.

The *Sullivan Principles* required companies doing business in South Africa to desegregate work facilities, provide equal employment practices for all employees, initiate training programs for minorities, promote minorities to management and supervisory positions, and improve the quality of employees' lives outside the work environment. In 1985 President Reagan signed an Executive Order that included a requirement that American businesses in South Africa abide by the *Sullivan Principles*. Neither Presidents Bush nor Clinton issued any executive order endorsing the *CERES Principles*; however, regional offices of the EPA have been supportive of CERES' effort.

CERES proponents believed that social investment forces could be marshalled to cause businesses to sign the *Valdez Principles* in much the same way as the multinational businesses signed the *Sullivan Principles*. CERES drafted its principles to address the release of pollutants, the sustainable use of natural resources, reduction and disposal of waste, energy efficiency, and conservation and risk reduction to employees and surrounding communities. CERES' objective was "to create a voluntary mechanism of corporate self-governance that will maintain business practices consistent with the goals of sustaining our fragile environment in future generations, within culture that respects all life ... " (Bavaria, 1989). CERES proponents believed that businesses who signed on would be recognized as leaders in making a voluntary public commitment to environmental accountability.

According to the CERES proponents, there are four advantages to businesses that agree to adopt the *Valdez Principles*. Companies can: (1) receive positive publicity that improves their corporate image in an age of "green consumerism;" (2) reduce waste-hauling fees and increase revenues generated by recycling; (3) strengthen their environmental standards and avoid financially devastating environmental disasters; and (4) receive investments from CERES members. CERES asked businesses to make a "quasi-legal commitment" by signing the Principles. Each year thereafter, signatory businesses would be required to submit detailed performance information setting forth how well they were complying with the mandate of the Principles. CERES planned to hire auditors annually to independently verify the businesses' self-audits and then grade each business against each Principle to provide an overall CERES score. CERES intended to disseminate the information as widely as possible, perhaps by publishing the environmental equivalent of a *Consumer Reports*. The original *Valdez Principles* stated:

VALDEZ PRINCIPLES

Principle 1: Protection of the Biosphere
We will minimize and strive to eliminate the release of any pollutant that may cause environmental damage to the air, water or earth or its inhabitants. We will safeguard habitats in rivers, lakes, wetlands, coastal zones and oceans and will minimize contributing to the greenhouse effect, depletion of the ozone layer, acid rain, or smog.

Principle 2: Sustainable Use of Natural Resources
We will make sustainable use of renewable resources, such as water, soils and forests. We will conserve non-renewable natural resources through efficient use and careful planning. We will protect wildlife habitat, open spaces and wilderness while preserving biodiversity.

Principle 3: Reduction and Disposal of Waste
We will minimize the creation of waste, especially hazardous waste, and wherever possible recycle materials. We will dispose of all wastes through safe and responsible methods.

Principle 4: Wise Use of Energy
We will make every effort to use environmentally safe and sustainable energy sources to meet our needs. We will invest in improved energy efficiency and conservation in our operations. We will maximize the energy efficiency of products we produce or sell.   *continued*

Principle 5: Risk Reduction
We will minimize the environmental, health and safety risks to our employees and the communities in which we operate by employing safe technologies and operating procedures and by being constantly prepared for emergencies.

Principle 6: Marketing of Safe Products and Services
We will sell products or services that minimize adverse environmental impacts and that are safe as consumers commonly use them. We will inform consumers of the environmental impacts of our products and services.

Principle 7: Damage Compensation
We will take responsibility for any harm we cause to the environment by making every effort to fully restore the environment and to compensate those persons who are adversely affected.

Principle 8: Disclosure
We will disclose to our employees and to the public incidents relating to our operations that cause environmental harm or pose safety or health hazards. We will disclose potential environmental, health or safety hazards posed by our operations, and we will not take any action against employees who report any condition that creates a danger to the environment or poses health and safety hazards.

Principle 9: Environmental Directors and Managers
At least one member of the Board of Directors will be a person qualified to represent environmental interests. We will commit management resources to implement these Principles, including the funding of an office of vice president for environmental affairs or an equivalent executive position, reporting directly to the CEO, to monitor and report upon our implementation efforts.

Principle 10: Assessment and Annual Audit
We will conduct and make public an annual self-evaluation of our progress in implementing these Principles and in complying with all applicable laws and regulations throughout our worldwide operations. We will work toward the timely creation of independent environmental audit procedures which we will complete annually and make available to the public.

The Response from Business in the U.S.

Unlike the *Sullivan Principles*, which gained fairly swift approval by multinational corporations, no major business in the U.S. signed the *Valdez Principles* when they were first announced. From September 1989 until April 1990, CERES solicited comments from more than 150 businesses on their views of the *Valdez Principles*. CERES also sent out approximately 3000 invitations to sign the *Valdez Principles*. This mailing included the Fortune 1000.

A common response was that businesses were disappointed that CERES did not request their input during the initial drafting process. Some objected to signing the Principles because they did not wish to subject their individual environmental and business practices to review and oversight by an independent special interest group. Nor did they wish their environmental performance to be audited and graded by environmental activists. One of CERES' important and noteworthy goals, however, was to actually build an environmental consensus among a broad spectrum of interest groups and to lessen the adversarial relationship between industry and environmental groups. This goal was not shared by the vast majority of companies that were invited to sign the Principles.

Nevertheless, some businesses were interested in reviewing the Principles and using them, as well as other information, to draft their own environmental policy statements without some of the more onerous public reporting requirements of CERES. Chevron dispatched representatives to talk with CERES and then, like many other major businesses, decided not to sign the Principles.

In April 1990 CERES issued a 22-page document entitled *The 1990 Guide to the Valdez Principles*. This guide was intended to address corporate concerns and identify particular actions a business would be required to perform if it chose to sign the Principles. Neither the guide nor persistent efforts by CERES members, however, resulted in a groundswell of corporate support for the *Valdez Principles*. CERES membership grew slowly to only 42 businesses — mostly small businesses and CERES members that had originally signed the Principles when they were first issued. Businesses that signed on included the Vermont ice cream company, Ben & Jerry's Homemade, Domino's Pizza Distribution, Inc., a lumber business, a paper recycling company, and two natural-ingredient, personal-care products businesses. Collectively, these business operations did not significantly impact the environment.

CERES proponents and other environmental activists who were dissatisfied with the corporate response increased social and political pressure on businesses to sign on. Shareholders' resolutions were filed with more than 50 corporations, urging them to adopt the *Valdez Principles*. Only 5 were voted on in 1990. That number increased to 31 in 1991. On May 24, 1991 at General Motors' annual meeting in Nashville, Tennessee, Chairman Robert C. Stempel argued against the business adopting such a resolution: "For more than three decades, General Motors has seen a clean and healthy environment as a top priority. We take pride in our own leadership role in reducing emissions from both vehicles and plants and in our work to minimize wastes and to dispose of those wastes in an environmentally sound manner." On the strength of these arguments, the GM resolution was defeated. By September 1992 similar resolutions failed at 56 other stockholders' meetings. The shareholder campaign has begun to dwindle over the last four years. From a record 54 resolutions filed in 1993, only 16 were submitted in 1996. According to the *Corporate Social Issues Reporter*, "[t]he eight resolutions that came to votes last proxy season were the lowest since 1990, the first year of the shareholder campaign."

Major businesses found several significant reasons not to adopt the *Valdez Principles*. Initially, businesses were troubled by the inclusion of a combination of ambiguous language and broad terms that might subject corporations to a bottomless snake pit of liabilities. They feared that legal actions could easily be taken against a business that has an unsuccessful environmental policy which states that it goes beyond existing laws. Some claimed that adopting the Principles might cause substantial and unjustified increased operating costs. Others felt they mandated disclosures that might interfere with the confidentiality of new product development.

Amoco expressed the views of many corporations that opposed being forced to appoint an "environmental director." It released a statement that "[s]pecial interest directorships, environmental or otherwise, are... bad policy from the standpoint of corporate governance." Some state legislatures, however, have actually endorsed the use of constituency directors by enacting statutes to allow corporate boards to be concerned with other interests, including the environment. Tom Smith, Vice President of Public Affairs for Dow Chemicals, also opposed environmental directorships and questioned: "Who is qualified to — quote 'represent environmental interests?'" John McCallister of DuPont added that board members must "represent stockholders, not constituencies."

Some businesses refused to sign the *Valdez Principles* because they believed that releasing extensive environmental reports to CERES might lead to undue outside influence and pressure on how they conducted internal environmental affairs. One activist predicted as much by advising CERES to establish reasonable benchmarks first so that organizers could recruit more businesses. He urged CERES to get businesses to sign on to the Principles and gradually increase the stringency of the standards. He argued that businesses would be reluctant to abandon the code once they have publicly signed.

CERES did not follow these bait and switch tactics. Instead, CERES thoughtfully listened to corporate complaints, modified some of the Principles, and changed the name from the Valdez to the *CERES Principles*. CERES eliminated the environmental director requirement and replaced it with the statement "[i]n selecting our Board of Directors, we will consider demonstrated environmental commitment as a factor." Also, a disclaimer was added to alleviate corporate concerns that the Principles would be used against endorsers in litigation: "[t]hese principles are not intended to create new legal liabilities, expand existing rights or obligations, waive legal defenses, or otherwise affect the legal position of any endorsing company, and are not intended to be used against an endorser in any legal proceeding for any purpose." Other revisions were made to make the Principles more palatable to businesses. For example, *Principle 5* ("We will minimize the environmental, health and safety risks to our employees and the community... by being constantly prepared") was revised to read, "We will strive to minimize... by being [ ] prepared." *Principle 7* ("We will take responsibility for any harm we cause to the environment by making every effort to fully restore the environment and to compensate those persons who are adversely affected") was changed to "We will promptly and responsibly correct conditions we have caused that endanger health, safety or the environment. To the extent feasible, we will address injuries we have caused to persons or damage we have caused to the environment and will restore the environment" (CERES Principles).

These significant revisions resulted in progress for CERES. On February 10, 1993 Sun Company, Inc., became the first Fortune 500 company to endorse the *CERES Principles*. Shortly thereafter, H.B. Fuller, a specialty chemical manufacturer, signed on — followed over the course of three years by General Motors, Polaroid, Bethlehem Steel, and BankAmerica.

After 8 years effort, however, only 50 businesses, including one Boston law firm, have endorsed the *CERES Principles*. Many businesses that decided not to endorse the Principles, nonetheless see the wisdom of implementing effective environmental management systems. CERES has been a positive influence in motivating businesses to adopt policies and evaluate and improve their environmental performance. Businesses have found it wiser and more productive, however, to design policy statements that fit their own individual needs, and advance their goals and objectives in the marketplace and the worldwide environment. Instead of adopting the mandatory self-audit and self-disclosure procedures designed by CERES, businesses have found their own way to present their environmental records to the public.

CERES' Impacts

It is hard to quantify CERES' overall impacts on corporate environmental performance, including self-reporting of environmental results. One of its proactive endorsers, Polaroid, produced one of the first environmental reports four years before it signed the *CERES Principles*. Recent statistics show that there is a growing trend to report. According to a study of U.S. businesses completed this year, approximately 120 large U.S. companies or 20% of selected industries from the S&P 500 and Fortune 500 issued a corporate environmental report by 1995 (Lober, 1997). Not all of these companies issue these reports annually. That same study found that 82% of the companies reporting indicated employees were one of their targeted audiences; the average number of reports printed is 28,000; over 20 companies are releasing their reports on the Internet; 40% of the reports include quantitative goals; 19% mention international releases; 17% describe legal actions pending; 13% include a third-party's assessment; 12% mention sustainability in their policy statement; and 12% compare their performance to competitors (Lober, 1997).

Undoubtedly, many businesses are preparing their own annual corporate environmental reports because they are proud of their environmental accomplishments. A recent international survey of 20 environmental reports (Elkington,

1996) reveals the following.

As expected, there is a huge diversity of different reports. All of the businesses surveyed, with the exception of one, are large. The reports range from an 8-page format to beyond 50 pages. Like the findings of the American study, very few reports benchmark their results against their competitors on a per unit-of-production basis. The exceptions according to the survey are Anglian Water's *1996 Environmental Report* and Shell U.K.'s *Environmental Report of 1995.* 40% of the businesses who issued environmental reports had their results verified by an independent third party; 12% offered their own verification statement. GM used its CERES report to demonstrate that its results were verified. Several businesses, including Volvo and Rohner, announced that they were considering verification in subsequent reports.

While environmental reporting is still in its infancy stage (see Chapter 7 for specific examples of environmental reports), it is remarkable how popular this method of benchmarking environmental performance has become in so short a period of time. CERES' request to companies to make a public disclosure of their environmental performance has had some effect on businesses deciding to become more proactive in disclosing their performance to the public. Other organizations have taken a more direct and confrontational approach to changing corporate environmental behavior.

CEP

Each year the Council on Economic Priorities (CEP) conducts a *Campaign for Cleaner Corporations (C-3)* to identify what it calls "some of the nation's worst corporate environmental offenders by comparative analysis of companies in the same industry" (Hilton and Marlin, 1996). In May of each year, CEP researchers commence an investigation of companies in a number of heavy industries, including petroleum, paper and forest products, electric utilities, automotive, chemical, and iron and steel manufacturing. CEP obtains information from the EPA's *Toxic Release Inventory* and other sources of information pertaining to disposal of wastes including the U.S. Coast Guard's *Emergency Response Notification System* (Hilton and Marlin, 1996). This information is evaluated by a panel of seven judges from academia, investment houses, public interest groups, and an environmental consultant. CEP researchers then rank companies in each industry according to their toxic releases. The rankings include data from the fugitive and stack air emissions, water and publicly owned treatment works, land and underground injection, hazardous waste generation, air compliance, and spilled hazardous materials.

By early June of each year, CEP notifies companies that they are under consideration for the campaign. Each company is sent an extensive questionnaire specific to its industry which covers its environmental policy, environmental impacts, and environmental litigation (Hilton and Marlin, 1996). CEP also requests a copy of each company's environmental policy and any information the company may wish to publicly disclose pertaining to environmental audits conducted by the company. CEP then offers each company the opportunity to respond, which may include supplying the company's own release figures and other environmental data. Finally, the judges meet to determine which company ranks the worst in its industry.

Approximately eight companies in different industries are then notified that they are on the C-3 list. Companies are given an opportunity to meet with the staff and judges to obtain practical environmental information as to how they can earn the right to be delisted. If companies significantly improve their environmental performance, they are eligible to be delisted the next year. Since its inception in 1992, all but three of the listed companies have been removed from the C-3 list. In 1995 CEP identified five companies that were delisted and the reasons why they were removed:

- **International Paper:** developed a comprehensive environmental report and achieved a 78% reduction in toxic releases tracked under the EPA's 33/50 Program based upon 1988 releases. It reduced its total toxic releases, agreed to change over its paper bleaching mills to be essentially chlorine free by the end of 1996, and is exploring closed-loop technologies.

- **Texaco:** signed an agreement with the government of Ecuador to remediate environmental damage caused by the company and agreed to spend $1 million in health facilities for Ecuadorian people living in affected communities. It pledged $1 million to finance natural resource projects, developed an environmental auditing program, and began to publish international data in its environmental report.

- **Union Carbide:** agreed to fund a hospital for victims of the 1984 Bhopal disaster and launched an effort to address its spills and accidents record, including a root cause analysis. It has had no worker or safety violations since being named to the campaign.

- **Westinghouse:** agreed to reduce toxic releases to below the average for the ten electrical equipment companies studied by CEP and made a commitment to adopt global environmental principles for all its facilities.

- **Westvaco:** The company announced a $140 million initiative to eliminate electrical chlorine in its paper bleaching mills, created a new corporate executive position (Vice President of Environmental Affairs), and achieved an 84% reduction in toxic releases since 1988 tracked under EPA's 33/50 program (Hilton and Marlin, 1996).

Most companies have worked with CEP cooperatively to become delisted. Several have challenged CEP's statistics, derived from governmental sources, and some have threatened litigation. Others have recognized positive gains from the process. For example, Lee Thomas, Senior Vice President of Environment and Governmental Affairs at Georgia Pacific, has stated that the dialogue between CEP and industries is important for continued environmental progress. Referring to his company, Thomas said, "We still have a lot of work to do, but your recommendations have helped us on issues where there is an opportunity to make even more progress" (Hilton and Marlin, 1996).

For most businesses, the chance of being placed on CEP's annual list is quite small. It appears that CEP is focused on the largest companies in heavy industry, although CEP adjusts its rankings for size by calculating emissions per $1,000 of sales. Being singled out by international, national, regional, or local organizations for a bad environmental performance record has adverse consequences for any business. In Colorado the media ranks the state's ten worst polluters each year. Whether the company's name appears in the *Denver Post* or CEP's list, the consequences are the same. Other organizations have attempted to focus on more positive ways to influence corporate performance.

NCPL

In 1994 Edward A. Dauer, the President of the National Center for Preventive Law (NCPL), brought together 33 experts from a broad variety of fields, including corporate risk management, law, education, and communications. Unlike CERES and CEP, these proponents of corporate compliance were not just interested in environmental protection. Motivated in part by the *U.S. Sentencing Guidelines'* directive that companies can reduce their exposures by adopting a compliance program, this group viewed their mission broadly — to assist companies to prevent and detect violations of any type of corporate law. The main areas of concern were antitrust and other fair trade laws, government procurement and contracting, political contributions and lobbying, securities and insider trading, money laundering, environmental issues, labor relations and employment discrimination, sexual harassment, intellectual property, substance abuse, product liability, consumer protection, workplace safety, conflicts of interest, and commercial bribery. Over the course of two years, this distinguished group of experts met and discussed how to achieve superior corporate compliance performance by creating effective compliance programs. The group's goal was to draft a unified and comprehensive set of principles and general guidelines for corporations to help them achieve compliance with civil and criminal laws.

In 1995 Dauer's group, formally known as the National Commission for Corporate Compliance, drafted 20 principles of compliance and numerous considerations that guide companies in implementing an effective compliance program. At the conclusion of the project, Dauer summed up the experience by stating that "corporate compliance is not achieved by assembling lists of *Thou Shalts* and *Thou Shalt Nots*. The *U.S. Sentencing Guidelines for Organizations* themselves are not a collection of crisp requisites, but rather a set of broad standards that define, in an open-textured and nondirective way, what counts as due diligence in the implementation of a system to 'prevent and detect violations of law'" (NCPL, 1996). Dauer added that the Commission's final product was not intended to establish standards or minimums, or even "best practices." The group simply created policies or activities that any size of company should consider whether to act upon or not as its own circumstances require. Hundreds of examples were added by the Commission's members to describe the innumerable ways that the principles and considerations can be put into action through a compliance program. While the Commission's focus was strictly on corporate compliance, which is a broader topic than environmental management systems, the principles, considerations, and examples are useful tools for companies to employ in managing and reducing environmental risk. The Commission's Principles are as follows:

NATIONAL CENTER FOR PREVENTIVE LAW CORPORATE COMPLIANCE PRINCIPLES

Establishing Compliance Programs

Principle 1

Organizations should pursue compliance through the creation and maintenance of an effective compliance program.

Principle 2

An effective compliance program is designed to prevent, detect, and respond to legal risks and to promote compliance with the law.

Principle 3

An effective compliance program is a dynamic process that is designed to be flexible and modified, when appropriate, to reflect changing conditions.

Principle 4

An effective compliance program states that it is the organization's policy to comply with all applicable laws.

Principle 5

The highest governing authority within an organization should endorse the organization's compliance program.

Principle 6

An effective compliance program establishes accountability for compliance throughout the organization.

continued

Principle 7
An effective compliance program is designed to operate fairly and equitably.

Structure and Control
Principle 8
Specific high-level personnel in an organization are responsible for the administration and oversight of the compliance program.

Principle 9
A compliance program has the support of senior management of the organization. Each officer, manager, and employee is responsible for supporting and complying with the compliance program's standards and procedures.

Principle 10
The organization exercises due diligence to prevent the delegation of substantial discretionary authority to persons having a propensity to engage in illegal activities.

Principle 11
The organization takes reasonable steps to achieve compliance with its standards and the law.

Principle 12
Incentives and disincentives are significant tools in promoting compliance.

Communications and Training
Principle 13
The organization's compliance program has a communications component, the objectives of which are to make employees and other agents aware of applicable standards of conduct and to promote compliance.

Principle 14
An effective compliance program communicates appropriate compliance information and motivation to the organization's employees and other agents.

continued

Principle 15
An effective communications program is designed to reach the intended audience.

Principle 16
All levels of management are responsible for the operation of an organization's compliance communications program.

<u>**Responses to Violations**</u>
Principle 17
An effective compliance program is proactive in its approach to dealing with incidents of noncompliance.

Principle 18
An effective compliance program possesses or has access to investigatory, evaluative, and reporting resources and utilizes those resources to monitor compliance.

Principle 19
An effective compliance program addresses the occasions for external reporting of violations of the law.

Principle 20
An effective compliance program utilizes incidents of non-compliance to evaluate its own effectiveness, to correct deficiencies, and to effect improvements.

Reprinted with permission from NCPL

THREE APPROACHES, ONE GOAL

CERES, CEP, and NCPL took three entirely different approaches to achieve the goal of improving corporate performance. All three organizations have been instrumental in providing guidance to companies in how to discover new ways to change existing corporate culture and conduct. Putting aside the advantages and disadvantages of each approach, all companies would agree that preserving corporate reputation is a significant reason to be proactive in this field.

A good corporate image, however, is only one reason to monitor and improve environmental performance. The satisfaction of saving money and minimizing the use of energy and natural resources by eliminating or diminishing environmental impacts is a powerful incentive for improving environmental performance. In the following chapters, there are a number of strategies to managing and reducing environmental risk. We focus initially on ISO 14000 because it presents a comprehensive systems approach to achieving better environmental results. Later chapters will discuss how such programs can be augmented by risk management and reduction strategies and initiatives. Far less confrontational than CERES or CEP, ISO 14000 adopts the same approach as the NCPL Commission, allowing businesses to select the pace and timing of improvements to environmental performance in a global marketplace that is increasingly becoming more interconnected and concerned with environmental degradation. Properly implemented, ISO 14000, as discussed in Chapter 5, can result in a wide range of benefits and opportunities for significant environmental and economic gains.

ISO 14000 presents another alternative for companies to deal with their rigorous environmental responsibilities. The next chapter discusses how the practice of environmental risk management has changed in the last quarter of the Twentieth Century, and how ISO 14000 can play a key role in the development of systems that will assure compliance with environmental laws and create opportunities to eliminate pollution and industrial waste in the future.

Chapter 3

NEW APPROACHES TO RISK MANAGEMENT SYSTEMS

Environmental risk management is finally coming of age. This new field has matured over the past several years through the incorporation of the diverse elements of communications, toxicology, insurance, litigation avoidance, and regulatory compliance. Until now, there has never been a coordinated, holistic approach to environmental risk management — one that integrates these seemingly disparate disciplines into a meaningful structure for senior management to make risk-based decisions. The ISO 14000 series of standards for environmental management presents a uniform approach. ISO 14000 attempts to solve the problems that have made environmental risk management more technical than strategic, more reactive than proactive.

RESPONSES OF COMPANIES TO THE REGULATORY ARENA

The U.S. Congress has passed a number of environmental laws designed to clean up past disposal and control the future disposal of pollutants and waste. Congress has delegated to the U.S. Environmental Protection Agency (EPA) the responsibility to promulgate regulations that achieve the broad legislative objectives of the *Clean Water Act*, the *Clean Air Act*, the *Safe Drinking Water Act*, the *Resource Conservation and Recovery Act* (RCRA), the *Comprehensive Environmental Response, Compensation and Liability Act* (CERCLA), and many others. The EPA, in turn, has issued many thousands of pages of complex regulations to protect the environment. No one is suggesting that all of these laws and regulations are flawed, or that they have not had beneficial effects for the country. Indeed, the EPA estimates that the health benefits of the *Clean Air Act* are between $10.5 trillion and $40.6 trillion at a cost to industry of $523 billion. It is the price we, as the public, pay for the regulations and their effectiveness that is in doubt. Many people believe that businesses have been too focused on responding to regulatory agencies, rather than on solving their own environmental problems.

Four out of five environmental rules created by the EPA are challenged in court. In the 1970s and 1980s, large companies created a whole separate environmental division under an environmental manager to respond to command and control direction from environmental authorities. These managers assumed the complicated and unenviable task of ensuring that their businesses complied with all the laws. Some believed that senior management should not be bothered by the details of environmental compliance. From the 1970s until recent times, many business owners and senior managers deferred to their environmental managers to handle the businesses' environmental matters and not to involve them unless major problems arose. Rigorous environmental regulations resulted in end-of-pipe solutions, and not effective management, in an effort to catch pollution at the end of the process. It has become obvious to people involved in the command-and-control regulatory framework that the current process was ineffective and unmanageable. Conducting business and government as usual is a costly and devisive method of protecting the environment.

CERCLA is one of the best examples of well-intended environmental legislation that has failed to achieve its goals. In 1980 Congress predicted that environmental remediation under CERCLA would ultimately cost several billion dollars. By 1989 the estimated cost for the remediation of hazardous waste sites had increased to $300 billion. Much of the money spent on these sites has gone to lawyers and consultants for transactional costs, including challenges to the scope and necessity of the clean-ups. Current estimates for the total cost of all environmental clean-ups in the U.S. range from $700 billion to well over $1 trillion. A consensus is building in the government, business, and the environmental community that to be most effective in their efforts to clean up and preserve the environment, businesses need to employ proactive management techniques to prevent contamination.

These efforts have been hampered by inadequate data generated by current, internal environmental management practices that rarely present a systematic picture of organizational operations, and in most cases is not adaptable to strategic decision making. Environmental data can be difficult to piece together to achieve an overall picture of where the business stands with regard to its environmental risks.

Environmental profiles and risk-related information are often pushed into the lower levels of management, where the sheer volume of data is difficult to interpret and manage. These data are generally not utilized to make better management decisions, nor are they considered in the allocation of scarce organizational resources. The end result is that environmental risks are generally ignored until they approach extreme levels, at which point senior management becomes involved. For many businesses, however, this is too late and their operational and financial performances suffer as they try to discover quick fixes for problems that have evolved over many years.

ISO 14000 SYSTEMS APPROACH

The ISO 14000 series of standards changes this scenario considerably. First, the standards mandate that environmental management is among the highest corporate strategic priorities. The standards provide a framework for top management to assess, manage, and reduce the risks associated with the environmental aspects of operations. Through several key requirements, the ISO 14000 standards, elevate environmental management into areas of strategic decision making, where, from an organizational perspective, more effective decisions for reducing risks can be made. Companies may no longer delegate environmental decision making only to low-level management personnel, who operate in an unchecked or unverified environment.

The ISO 14001 standard for environmental management outlines an organizational framework for the systematic identification, control, and improvement of all environmental *aspects* and *impacts*. ISO 14001 defines an environmental aspect as an "element of an organization's activities, products or services that can interact with the environment." An environmental impact is "any change to the environment, whether adverse or beneficial, wholly or partially resulting from an organization's activities, products or services." ISO 14001 requires the development of a plan to get management and employees to commit to the protection of the environment with clear assignment of accountability and responsibility.

Rather than focusing exclusively on environmental performance, the standard addresses the management systems used to control environmental performance. This is a marked departure from the command-and-control approach currently utilized by state and federal environmental protection agencies to regulate business activities. Management is still accountable for attaining emission levels prescribed by regulations and agreements. The integrated series of management systems in ISO 14000, however, motivate and encourage management to actively seek ways to reduce or eliminate pollutants and go beyond compliance.

The ISO 14000 standards are the result of, and driven by, four ideological views: (1) existing environmental management practices are generally inadequate and ineffective; (2) businesses desire to improve environmental performance for economic and social reasons; (3) regulators, stakeholders, and the public will increasingly hold businesses responsible for their impacts to the environment; and (4) environmental management systems must be an integrated and efficient part of the business processes. With these viewpoints in mind, ISO 14001 management system implementation can require significant and profound change.

The ISO 14001 standard for environmental management systems requires five basic and logical activities that businesses must follow:

- Establish senior management commitment to environmental management and promulgate a comprehensive environmental policy.
- Develop objectives, targets, and a program to implement the environmental priorities stated in the policy.
- Perform the activities necessary to achieve the objectives and targets, develop documents and records, and train employees in their environmental responsibilities.
- Monitor and measure on a regular basis the performance of the environmental management system.

- Review the entire set of environmental management activities periodically to ensure continual improvement.

These five basic activities are intended to facilitate the allocation of resources; the assignment of responsibilities; and the ongoing evaluation of practices, procedures, and processes to ensure achievement of the environmental policy requirements. Thus, the ISO 14001 standard views the business from the process perspective; defines the boundaries of environmental responsibility through the policy statement; and seeks structural solutions through comprehensive monitoring, internal audits, and management reviews.

ISO 14001 recognizes the importance that governments, industry leaders, and stakeholders work together to design and implement administrative and structural procedures to improve environmental conditions. Most businesses will have to modify their corporate behavior to implement the standard. In particular, some will have to learn how to develop better working relationships with the public and engage in collaborative decision making to improve environmental performance. Governments will need to redirect regulatory agencies to change existing regulatory schemes and create further incentives for business to protect the environment. The federal and state governments will need to develop better working relationships with each other so that they no longer compete and interfere with each other's statutory responsibilities, as changes occur in the manner that businesses self-regulate and report to these entities.

With these global changes needed, skeptics may find the required tasks overwhelming. The evolution of these changes, however, is already beginning. In the U.S., the EPA is developing many programs designed to allow businesses more latitude to explore ISO 14000 and learn how to self-regulate and self-disclose their environmental performance before unnecessary problems escalate into environmental noncompliance. Chapter 14 describes these new EPA initiatives, and Chapter 11 sets forth how businesses can take advantage of federal and state programs designed to allow businesses to self-disclose environmental problems and take effective corrective action. These programs can often significantly reduce or eliminate penalties for offenses that occur. The U.S. Department of Energy (DOE), under its General Environmental Protection Program, is adopting the ISO 14000 approach and developing its own environmental management systems. Throughout Europe and Asia, government bodies are working with industry leaders to implement the requirements of ISO 14001.

ISO 14001 provides businesses with a practical and workable framework for managing environmental risk. Its focus on continual improvement and prevention of pollution encourages businesses to move from risk financing into comprehensive risk management activities. It also ensures that a business adopts the process perspective and systematically evaluates and analyzes existing and potential exposures before losses occur. With ISO 14001, environmental decisions become strategic concerns, where risks can be assessed and long-term resource allocations can be made — considering both external pressures and operational priorities.

Implementation of a comprehensive and integrated environmental management system has the effect of simultaneously managing and reducing environmental risk. These two separate but interrelated concepts are merged in the systems approach. ISO 14001 requires businesses to determine the legal and other requirements and environmental aspects associated with the businesses' activities, products, and services. A process must be defined by businesses to achieve targeted performance levels and engage in environmental planning throughout the product or process life cycle. Companies are required to provide appropriate and sufficient resources to achieve targeted enforcement levels on an ongoing basis, while establishing and maintaining communications with internal and external interested parties. Appropriate and sufficient resources must also be allocated for training to achieve targeted performance levels on an ongoing basis. Finally, ISO 14001 provides management with a process to audit and review the environmental management system and to identify opportunities for improvement of the system and the resulting environmental performance.

By effectively implementing ISO 14001, senior management ensures that it is managing its environmental risks and limiting its exposures. Structural solutions are addressed and the need for risk financing is reduced. The environmental loss controls inherent in the ISO 14000 management systems also place an entity in a better position to access environmental insurance or other risk financing methods for an acceptable cost. ISO 14000 is a systems framework, enabling businesses to include environmental risk management as a strategy for their future. The ISO 14000 environmental management system coupled with other risk management and risk reduction techniques comprises the systems approach. The next chapter explores in more detail how ISO 14000 works and provides an overview for businesses to begin to consider how the standard can be applied to their systems operations.

Chapter 4

MANAGING RISK USING ISO 14001: A GENERAL OVERVIEW

Chapter 3 introduced the five basic ISO 14001 organizational activities. Senior management must: (1) commit to proactive environmental management and promulgate a comprehensive environmental policy; (2) develop objectives, targets, and a program to implement the environmental priorities stated in the policy; (3) perform activities necessary to achieve the objectives and targets, develop documents and records, and train employees in their environmental responsibilities; (4) monitor and measure on a regular basis the performance of the environmental management system; and (5) review the entire set of environmental management activities periodically to ensure continual improvement. Within those organizational activities there are 17 different environmental management functions, indicated in the box below. This chapter will illustrate management activities within the framework of ISO 14001 that are organized and implemented to create an effective environmental management system.

ISO 14001 ORGANIZATIONAL ACTIVITIES AND ENVIRONMENTAL MANAGEMENT SYSTEMS

Commitment and Policy
1. Environmental Policy

Planning
2. Environmental Aspects
3. Legal and Other Requirements
4. Objectives and Targets
5. Environmental Program

Implementation
6. Structure and Responsibility
7. Training, Awareness, and Competence
8. Communications
9. Environmental Management System Documentation
10. Document Control

continued

---

11. Operational Control
12. Emergency Preparedness and Response

Monitoring and Measuring
13. Monitoring and Measurement
14. Nonconformance and Corrective and Preventive
    Action
15. Records Management
16. Environmental Management Systems Audits

Review and Improve
17. Management Review

ESTABLISHING A RISK BASELINE USING THE INITIAL REVIEW PROCESS

The first element of effective risk management is to undertake a systematic review of a business's operations to identify risks. ISO 14001 requires a business to conduct an extensive assessment of all operations that have the potential to impact the environment and of the systems that control the identified risks.

The initial review covers four basic areas: (1) identification of environmental aspects of the business's activities, products, or services to identify those that have been or can have significant environmental impacts and liabilities; (2) identification of regulatory and other requirements; (3) an examination of all existing environmental management practices and procedures; and (4) an evaluation of feedback from the investigation of previous incidents.

The company must first conduct an environmental aspects analysis to identify the potential impacts of its activities, products, or services. The business must consider these aspects and how they are related to the potential impacts to the environment in setting its environmental objectives. Significant impacts include emissions to air; releases to water; waste management; contamination of land; use of raw materials, energy, and natural resources; and any other activities that impact the environment.

After the organization has identified its environmental aspects and impacts, it must analyze the consequence of potential impacts through a review of pertinent regulations and other requirements. A review of the business's legal obligations is necessary to rate the severity of potential impacts to the environment.

The initial review should include an assessment of the existing environmental management practices and procedures, current regulatory posture, codes of conduct or practices, and sets of principles and guidelines used by the company or its industrial counterparts. The initial review should also identify existing policies and procedures dealing with procurement and contracting activities. Opportunities for competitive advantage, the views of interested parties, and feedback from the investigation of previous incidents of noncompliance need to be included in the initial review. Prior performance of the company can be evaluated and compared with relevant internal and external standards. Functions or activities of other organizational systems can be analyzed to determine whether they impede or improve the likelihood of enhanced environmental performance.

At the conclusion of the initial review, the results should provide senior management with a complete picture of organizational processes and their associated environmental impacts. The initial review will serve as a baseline of environmental exposures from which a performance plan with targets and objectives can be established. In one document, senior management should have a clear understanding of what operations are being performed, what environmental aspects of those operations are most likely to impact the environment, what regulatory requirements are applicable to each, and where current or potential risks exist. With this information, senior management can thus begin the task of preparing an environmental policy statement for the company.

ENVIRONMENTAL POLICY

In the ISO 14001 process, the results of the initial review are translated into a policy statement, identifying objectives and targets, and establishing an environmental program as the backbone of the risk management plan. The policy provides the impetus for a series of actions to minimize risks emanating from uncontrolled emissions, noncompliance with regulatory requirements, inadequate environmental management practices, and uncorrected adverse conditions. This is also the opportunity to further reduce environmental risks by removing or minimizing the causes of pollution, and any remaining liabilities from previous incidents.

ISO 14001 requires top management to develop a corporate environmental policy statement and to ensure that it:

- Is appropriate to the nature, scale, and environmental impacts of its activities, products, or services
- Includes a commitment to continual improvement and prevention of pollution
- Includes a commitment to comply with relevant environmental legislation and regulations, and with other requirements to which the organization subscribes
- Provides the framework for setting and reviewing environmental objectives and targets
- Is documented, implemented, maintained, and communicated to all employees
- Is available to the public

The corporate environmental policy statement is the driver for implementing a new environmental management system and the company can use the policy and the system to maintain and improve its environmental performance. The policy "establishes an overall sense of direction and sets the principles of action for the organization" (ISO 14004). The ISO 14001 authors were well aware of the CERES principles, the International Chamber of Commerce's guidelines, the chemical industry's Responsible Care® program, and other industries that have developed individual policies and programs. These guidelines formed a basis for defining the necessary commitment to the environment through the environmental policy statement and the common set of values needed to produce an international standard for environmental management systems.

Effective environmental policies cover areas of high risk for the business. Specific environmental policies from a wide range of businesses, which were prepared before ISO 14000, are reviewed in Chapter 10. Many of the concepts contained in these policy statements found their way into ISO 14000. In addition to the required elements presented above, an environmental policy may also address the following environmental risks facing a business:

- Minimizing risk to employees and communities
- Emergency preparedness
- Defining responsibility and authority for environmental management
- Providing employee education and awareness
- Evaluating the environmental impact of processes and materials
- Reducing, recycling, and reusing materials
- Minimizing waste and reducing emissions
- Cooperating with regulatory agencies
- Actively participating in public policy development
- Measuring attainment of objectives and goals

Management should use the environmental policy statement to guide the business in its quest to reduce or eliminate risks based on the integration of accurate information regarding organizational processes and overall environmental objectives and priorities. By following up with a management review of the policy at periodic intervals, the standard assures that environmental information becomes the purview of senior management. Environmental performance is thus included in the strategic decision-making process. In addition, compliance with regulatory requirements becomes only one aspect of risk management and going beyond compliance becomes the accepted way of doing business.

ENVIRONMENTAL ASPECTS

The formal preparation of an environmental management system requires the business to address all aspects of its activities, products, or services that can impact the environment. As previously discussed, an environmental aspects analysis must be conducted to identify environmental aspects of activities, products, and services under the control or influence of the business. A formal procedure must be instituted providing how the business will conduct the aspects analysis. The analysis must determine which aspects have significant impacts on the environment. This analysis must both identify and rate the potential severity of the identified aspects and impacts on the environment. The standard also requires that the information must be kept up-to-date. The identified impacts must then be considered when developing the corporate environmental objectives. In reality, the environmental aspects and impacts analysis is the initial site-specific data collection phase of the environmental management system, and should be conducted prior to the development of the environmental policy statement.

LEGAL AND OTHER REQUIREMENTS

A formal procedure is required to identify the legal and other requirements that are applicable to the environmental aspects of the business. The standard requires identification not only of governmental regulations, but also of industry associations or groups, commercial standards of practice, and professional codes. These requirements should be readily accessible, updated, and available to employees.

OBJECTIVES AND TARGETS

The standard requires that there be documented environmental objectives and targets at each relevant function and level within the business. The objectives and targets should consider legal and other requirements; significant environmental aspects; technological options; financial, operational, and business requirements; and the views of interested parties. The corporate objectives and targets should be consistent with the corporate policy. They should include commitments such as pollution prevention and improvements in uses of energy and natural resources, as well as continual improvement of system operations.

ENVIRONMENTAL MANAGEMENT PROGRAMS

The standard requires that there be programs instituted for achieving the identified objectives and targets. These programs must define the means, schedule, and responsibility at each relevant function and level to achieve the objectives and targets. The programs should be able to be amended as necessary to include new developments or modified activities, products, or services — yet still attain the objectives and targets.

STRUCTURE AND RESPONSIBILITY

ISO 14001 specifies that the roles, responsibilities, and authorities shall be defined, documented, and communicated in order to facilitate effective environmental management. Management needs to provide human resources with specialized skills, technology, and financial means to implement and control the environmental management system. Top management must appoint specific management representatives who have defined roles, responsibility, and authority for ensuring that the environmental management system is established, implemented, and maintained in accordance with the standard. There must also be a reporting system on the performance of the environmental management system to top management for review as a basis for improvement of the system. In large or complex organizations, ISO envisions that there may be more than one designated management representative. In small- to mid-sized enterprises these responsibilities may be undertaken by only one person. Top management must also ensure that sufficient resources are obtained to implement the system and that key environmental responsibilities are well defined and communicated to the relevant personnel.

Deployment of responsibility for environmental impacts into the business is very important. Many businesses have already achieved deployment with the ISO 9000 quality management systems standard. That standard embraces the principle of making product and service quality the responsibility of those performing the work. By including environmental requirements in each job, management makes all employees responsible for potential emissions to the environment, and supports them by providing resources to control and eliminate these emissions. Risks of unplanned or accidental emissions are reduced as emissions are routinely controlled by those directly involved in the operations, rather than by external personnel in an oversight control method, such as compliance auditing.

ISO 14001 promotes the idea that minimizing environmental impacts is the responsibility of each employee. An organization-wide awareness is created, which in turn is supported by comprehensive training programs. These actions greatly reduce the risk of unplanned emissions, discharges, or other environmental impacts, as well as the risk that an employee will violate the law, either knowingly or unknowingly.

TRAINING, AWARENESS, AND COMPETENCE

Training is another important function of the environmental management system. The organization is required to set up a structure that will identify training needs. Those employees whose activities significantly impact the environment must receive adequate training. Procedures must be established and maintained to make employees at each relevant function and level aware of:

- The importance of conformance with the environmental policy and procedures and with the requirements of the environmental management system

- The significant environmental impacts, actual or potential, of their work activities and the environmental benefits of improved personnel performance

- Their roles and responsibilities in achieving conformance with the environmental policy and procedures and with the requirements of the environmental management system, including emergency preparedness and response requirements

- The potential consequences of departure from specified operating procedures

The training provision further provides that personnel who perform the tasks which can cause significant environmental impacts must be competent and have the requisite degree of education and training and/or experience to do their jobs. The business needs to establish and maintain procedures for identifying the training needs. A business also should require contractors working on its behalf to demonstrate that their employees have the requisite training.

When employees are properly trained, there is less likelihood of noncompliance as a result of employee error. Employees must be given adequate resources and training to implement and control their aspects of the environmental management system and to ensure that they are competent and informed about the performance of the system. These actions can serve to increase the employees' satisfaction levels and reduce the likelihood of noncompliance or their premature reporting to the regulatory authorities.

COMMUNICATIONS

ISO 14001 requires that the business shall establish and maintain procedures for internal communications between the various levels and functions of the organization. The standard also requires procedures for receiving, documenting, and responding to relevant communications from external interested parties. This procedure can include a dialogue with interested parties outside the business enterprise and consideration of their relevant concerns.

When they are effectively implemented, the ISO 14001 requirements for internal and external communications create a positive perception of the business and its environmental policies. A dialogue with stakeholders can help organizations deal with recurring or emerging issues of concern and demonstrate a proactive stance on the environment. Communications open the door to collaborative solutions to environmental problems before they become disputes. Open channels of communications, coupled with a sincere commitment to solving problems, reduce the risk of citizens' suits and negative public and customer perceptions. A business needs breathing room when resolving a contentious or sensitive environmental issue.

Management can use its communications procedures to discuss its environmental impacts with diverse groups in a wide variety of situations. It can ensure that community perceptions of environmental risks are accurate and undistorted by rumor or the media. In some cases, a business may even be able to demonstrate that its environmental protection activities actually save a community money or maintain and create jobs, both of which cast a positive light on a business and its management.

ENVIRONMENTAL MANAGEMENT SYSTEM DOCUMENTATION

Documents and records are another important part of risk management. ISO 14001 provides that the business shall establish and maintain information in paper or electronic form to describe the core elements of the management system and to provide direction to related documentation. By requiring comprehensive document control of the environmental management system, the ISO 14001 standard provides for the gathering and retention of essential information about the environmental performance of the business. The level of detail has to be sufficient to describe the core elements of the environmental management system and their interaction. It must provide direction on where to obtain more detailed information on the operation of specific parts of the system. This information may be crucial in collecting documentation regarding the position of the business should any dispute arise, or should the business decide to report its environmental performance to interested parties. Most businesses will find that as they refine and centralize their document and records management systems, and duplication and redundancy are reduced, important data will become more accessible.

DOCUMENT CONTROL

The ISO 14001 standard requires the development and implementation of a logical system of document control. The requirements for document control enable management to locate and retrieve the documents necessary for operating the systems, thereby managing operations and reducing risks of uncontrolled events. The documents essential for the operation of the environmental management system need to be periodically reviewed; current versions must be available where operations essential to the effective functioning of the environmental management system are performed; obsolete documents must be removed; and documents must be retained and identified for legal and/or knowledge preservation purposes. The documentation may be integrated with documentation from other systems. Chapter 16 describes how a document control system can be created to reduce environmental risk and liabilities.

OPERATIONAL CONTROL

To further manage existing risks, the standard requires a comprehensive definition and implementation of operational controls to ensure achievement of objectives and targets. The controls provide direct written assurance that operations have been evaluated and are being carried out under optimum conditions. The controls should also ensure that suppliers and contractors are involved in responsible environmental management.

OPERATIONAL CONTROLS INCLUDE:

- Establishing and maintaining documented procedures to cover situations where the absence of procedures could lead to deviations from the environmental policy, targets, and objectives
- Stipulating operating criteria in the procedures
- Establishing and maintaining procedures related to the identifiable significant environmental aspects of goods and services used by the organizations
- Communicating relevant procedures and requirements to suppliers and contractors

EMERGENCY PREPAREDNESS AND RESPONSE

The 14001 requirements for emergency preparedness and response can assure managers that contingency planning and crisis management are integral parts of the business. The standard provides that a business will establish and maintain procedures to identify and respond to accidents and emergency situations. It also provides procedures to prevent and mitigate the environmental impacts that may be associated with them. The standard thus ensures that if the emergency preparedness and response procedures are effectively implemented, any potential impacts resulting from accidents will be dealt with in an appropriate manner.

Stakeholders can be assured that through crisis management the business has planned for unexpected events and is acting as a responsible corporate citizen. The likelihood of citizens' suits or poor community relations are effectively minimized. In developing these procedures, management will greatly reduce the risk of accidents as it considers the range of probable emergencies and their impacts, and prepares for contingencies in advance of their occurrence.

MONITORING AND MEASURING

Another area that can improve risk management practices is in the generation of timely and accurate monitoring and measuring data involving the environmental aspects and impacts of operations. The standard requires routine monitoring and measuring of environmental performance of the characteristics of its operations and activities that can have a significant impact on the environment. Information to track performance must be maintained as well as records of the calibration and maintenance of the monitoring equipment. The business must also establish and maintain a documented procedure for periodically evaluating compliance with relevant environmental legislation and regulations.

NONCONFORMANCE AND CORRECTIVE AND PREVENTIVE ACTION

Procedures must be implemented to define responsibility and authority for handling and investigating nonconformance, for taking action to mitigate any impacts caused, and for initiating and completing corrective and preventive action.

Internal auditing provides management with data about the degree of implementation and effectiveness of the management systems used to achieve environmental performance. These types of data can be tracked, trended, and used to enhance the organizational learning process, and can serve as the basis for continual identification and reduction of risks. The ISO 14001 standard fosters better management decisions and a clearer basis for risk reduction because it improves the quality of the environmental performance information available to management.

RECORDS MANAGEMENT

The ISO 14001 standard requires the development and implementation of a logical system for the identification, maintenance, and disposition of environmental records. These records will include training records and the results of audits and performance reviews. The requirements for document control ensure that management can locate and retrieve the documents which are protected against damage, deterioration, or loss. Their retention times must be established and recorded.

ENVIRONMENTAL MANAGEMENT SYSTEM AUDIT

The standard also provides for the business to establish and maintain programs and procedures for periodic environmental management systems audits to determine whether the system conforms to the planned arrangements for environmental management and whether it has been properly implemented and maintained. The audit program must be based on the environmental importance of the activities concerned and the results of the previous audits. The procedures must cover the audit scope, frequency, and methodologies as well as the responsibilities and requirements for conducting audits and reporting results.

The routine environmental audits required by ISO 14001 should identify any existing gaps in knowledge and data that managers must consider as they define and control risks. (Chapter 17 describes in more detail the gaps analysis process.) One of the biggest benefits of an ISO 14001 environmental management system is the focus on filling in these data gaps in a comprehensive manner. By ensuring that environmental data come from a broad array of activities related to specific objectives and targets, management will have high-level data integrity for strategic environmental decisions.

MANAGEMENT REVIEW

Finally, the standard contains a provision for management review of the entire system, in light of actual performance and external conditions. This provision enables an ongoing strategic risk management process, in which both internal and external changes in risk exposures will be fully considered and included in strategic decisions. The management review also provides the foundation for continual improvement of the system.

The management review should include:

- Results from audits
- The extent to which objectives and targets have been met
- The continuing suitability of the environmental management system in relation to changing conditions and information
- Concerns among relevant interested parties

The management review engages strategic planners and senior management in the task of evaluating the effectiveness of existing systems, and seeking structural solutions for reducing risks and preventing undesirable incidents. This review must consider implicit and explicit issues; intended and desired results; details of management actions; and changes in the organization, the marketplace, government, and communities during the last planning period. The combination of intended and actual results is called realized strategy. All observations, conclusions, and recommendations need to be documented for necessary actions. The management review allows merging of environmental considerations with financial objectives, customer satisfaction criteria, measures of internal management excellence, and community activities.

During the review process, senior management should address the implicit and explicit aspects of the business to take advantage of the opportunity to build customer satisfaction. Management may also decide to enact other new or related actions, such as product labeling and redesign, removal of toxic or hazardous materials from production processes, elimination of packaging, or development of the infrastructure for product recycling purposes.

The ISO 14001 standard can serve as an excellent organizational risk management framework in which a business considers, plans for, and manages environmental aspects and impacts before extreme conditions exist. By setting up systems that conform to the standard, senior management will be confident that it has handled its environmental risks and limited its exposures in a systematic and controlled manner. In ISO 14001, environmental decisions are strategic — risks can be assessed and long-term resource allocations can be made in light of competitive considerations and existing operational priorities.

Assessing Risks

Management often receives conflicting advice regarding the risks associated with operations and properties. It is thus important that managers have an adequate framework to accurately assess and prioritize environmental risks. ISO 14000 provides a structure to manage and understand various environmental risks facing a business.

Environmental risk traditionally has included historic liability risk, operating risk, marketplace risk, capital cost risk, transaction risk and sustainability. Although the historic liability risks identified during an environmental site assessment are becoming quantifiable, little hard data are available regarding the assessment and management of ongoing operational risk.

Environmental risk quantification differs from quantification of other casualty risks because of the relatively small amount of historical data, and the complexity of the contamination distribution variables. Ten years of worker's compensation loss experience, for example, is useful information upon which to base future loss projections, but ten years of soil contamination data are not likely to accurately predict future liabilities at a given site. Numerous variables, such as tank construction and installation, stored contents, local hydrogeology and geology, surrounding populations, and applicable laws and regulations, make it difficult to predict future environmental outcomes. In the absence of adequate historical data, the management systems, or engineering controls, must be used to develop an accurate assessment of environmental risk.

Risk Modeling

Experts in environmental, insurance, and risk management firms have developed risk modeling techniques to be used in the ongoing assessment of environmental risk. After the business has performed its initial review of its environmental aspects and impacts, legal and other requirements, and obtained its risk baseline, those data can be used in risk modeling to plan and prioritize.

Generally, risk modeling is a method of breaking down a complex situation into component parts, arranging those parts into hierarchical order, assigning numerical values to subjective judgments regarding the relative importance of each variable, and synthesizing the judgments to determine which variables have the highest priority. When using these models effectively, a business can act upon the highest risk concerns to predict and prevent loss, thereby controlling risk.

ISO 14000 can take the results from these risk assessing and risk modeling processes and use them in developing a comprehensive environmental management system that properly manages and ultimately reduces these risks. The benefits of developing a systems approach using as its centerpiece an ISO 14001 compliant environmental management system are set forth in the next chapter.

Chapter 5

BENEFITS AND COSTS OF DEVELOPING ENVIRONMENTAL RISK MANAGEMENT SYSTEMS

Management should consider the benefits of developing and implementing an environmental management system from six perspectives: economic, social, political, technological, ideological, and financial. Opportunities and risks to a business can arise in any or all of these areas. By reviewing each perspective, the problem is viewed in a macro context, where more meaningful evaluations of the business's operations and management systems can be ascertained.

ECONOMIC BENEFITS

The economic sector consists of the global, national, and local conditions of production, distribution, and service. Included are consumer spending, inflation rates, interest rates, labor supply, cost and availability of natural resources, financing methods, industry performance ratios, business investment patterns, and other conditions of production and competition.

Adopting an ISO 14001 environmental management system enables a business to achieve worldwide qualified supplier status, which can result in an expanded customer base. When the business is placed on a registry of certified businesses, it can be considered for contracts requiring certification as a prerequisite. Certification thus becomes a means of responding to changing market pressures to maintain and expand the current customer base.

As with the ISO 9000 series of quality management standards, the markets for ISO 14001 are expected to experience cascading certification prerequisites beginning with large multinational businesses, and those dealing with the U.S. Departments of Energy and Defense. First- and second-level suppliers to these large businesses may find that obtaining an ISO 14001 certification is a prerequisite for doing business with their existing customers, and that these markets may become closed if certification is not obtained. As numerous businesses obtain their ISO 14001 certifications, increasing pressure will also be placed on third- and fourth-level suppliers to become certified.

Internal operating efficiencies and cost reductions are also benefits to the business. By performing a process analysis and defining in detail operating processes that impact the environment, management also has the opportunity to reduce duplicative efforts and eliminate redundant systems. Very often the exercise of process mapping reveals many convoluted processes that have developed over time without being rationalized, evaluated, or balanced with other similar processes.

Process redesign, which identifies and isolates polluting processes, can assist management in reducing waste and energy usage. Pollution prevention, waste minimization, substitution of less toxic materials, and design for the environment are techniques management can use to reengineer existing processes to reduce or eliminate waste and polluting by-products. Incorporating a life-cycle perspective allows management to extend its control over environmental impacts beyond the facility and to address issues such as the types of raw material streams used, transportation of raw and processed materials, and product disposal. These issues, when properly addressed, can translate into operating efficiencies and cost reduction regardless of life-cycle stage in which the business operates. The life-cycle perspective also allows management to look at the possibility of recovering costs through recycling of packaging and manufacturing wastes. This cost recovery includes facility operations, as well as upstream operations such as raw material extraction, and downstream operations such as product use.

A major area of interest for senior management is to decrease liability exposure from environmental incidents and accidents. By identifying high-risk processes and working to reduce their environmental impacts, management can address areas of concern and ensure that risks are systematically reduced or eliminated. These efforts should be detailed in an environmental management system, so the business can obtain cost reductions through reduced insurance rates and access to capital at lower-than-market rates. The recent Federal Deposit Insurance Corporation guidelines for lenders also prescribe activities that evaluate the environmental management practices of potential borrowers. The Environmental Bankers Association, with 50 member-institutions worldwide, has taken this information and is developing a clearinghouse of information from the government and its member-institutions concerning the environmental practices and procedures of the lending industry. The EPA's Merit Program, described in more detail in Chapter 14, is addressing how ISO 14000 can be used to reduce the cost of capital. By implementing an ISO 14001 environmental management system, a business can demonstrate to lenders that it meets or exceeds accepted lending standards in all respects, thus ensuring access to capital and maintaining positive relations with lending institutions.

Future court decisions may consider an ISO 14001 certification as evidence of a good environmental management system for purposes of penalty mitigation. Fines imposed on a business as a result of violations of environmental law can be adjusted downwards within a wide range according to the *Federal Sentencing Guidelines for Organizations*. These guidelines allow for establishing a range of fines based on six culpability criteria. The criteria include the level of authority and size of organization; prior history; violation of an order; obstruction of justice; an effective program to prevent and detect violations of law; and self-reporting, cooperation, and acceptance of responsibility. The guidelines allow for either the reduction of EPA penalties for a business with an environmental management system in place or an increase in EPA penalties for a business that does not demonstrate responsible environmental management. The implementation of an ISO 14001 environmental management system addresses several of these responsible management criteria to allow for mitigation of fines should a violation occur.

Large businesses are frequently faced with a number of different customer, regulatory, and registrar audits of their operations. Those doing business internationally are also often required to perform multiple inspections, certifications, and product registrations in order to demonstrate conformance with a varying array of regulations, requirements, and other technical specifications. The activities required under an ISO 14001 environmental management system provide consistent performance data and rationalized processes that should allow management to reduce the number of audits and inspections. This reduction translates directly into cost savings for the business.

SOCIAL BENEFITS

The social arena is focused on people, communities, and society at large. It includes population characteristics and trends, lifestyles, values, ethical standards, attitudes, public opinion, educational patterns, social change movements, and nonprofit groups and organizations. When a business adopts an environmental management system, the social perspective allows management to see how well the system can integrate the business into society at large.

Over the past twenty years, the growing awareness of the environment as an important issue has led to an emergence of an opportunity for businesses with environmental impacts. A business can project a socially responsible image by integrating an environmental management system into its operations. To ensure that the business's communication efforts will not be mistaken for *greenwashing,* the use of national and international standards can lend much credibility to the business's claims. Those standards provide a consensus approach that has been debated and agreed to by experts from more than 111 countries. Adoption of standards such as ISO 14001 enables a business to demonstrate a sincere and credible commitment to the environment, and to base its claims on a system that represents the state-of-the-art worldwide.

As business leaders begin to broaden their perspectives, they realize a wide array of stakeholders or interested parties are affected by and concerned with their operations. The growing trends of environmental activism, heightened awareness of environmental impacts due to increased reporting, and decreasing acceptance of businesses seen as environmental bad actors provide an opportunity for businesses to satisfy stakeholder interests for corporate accountability. A business can promote environmental awareness and ecological responsibility as cornerstones of its operating principles by the development and implementation of a good corporate citizen policy as the basis for its environmental management system. By using a communications-based framework to demonstrate its commitment to these principles, management can embed environmental excellence into formal and informal reports, town meetings, and product and services descriptions.

A key aspect of any ISO 14001 environmental management system is the integration of environmental issues into strategic decisions. By proactively addressing existing trends and emerging societal issues, a business can ensure that stakeholder perceptions of the risks posed by the business are accurate. Management can be assured that it is monitoring and addressing issues and trends that could have an adverse impact on the business by the continual process of strategically reviewing the environmental management system. Proactive management also ensures that management is holding itself accountable to the wide variety of stakeholder interests and concerns.

Customer relations can be positively affected by ISO 14001 activities. Customers can gain assurance that the business will not be shut down or dissolved due to environmental incidents or accidents. Customers, end producers, and contractors all gain more confidence in the integrity of management and are more likely to continue and extend their relationships when a business takes steps to ensure an uninterrupted supply of products and services.

How a business is performing from its customers' perspective has become a priority for well-run businesses. It is imperative that a business translate its general mission statement on customer service into specific environmental objectives and measures that reflect the concerns truly important to customers — cost, performance, quality, service, and time.

Although management may enact controls for minimizing the release of a certain chemical, the unintended use or disposal of products may have a severe effect on customer perception of business as being environmentally responsible. During the management review, senior management should address the implicit and explicit aspects of this issue to take advantage of the opportunity to build customer satisfaction. Management systems can identify the financial reasons and environmental consequences of product labeling, product redesign, awareness and elimination of toxic materials, and packaging or development of infrastructure for product reuse and recycling. Innovations made possible by top management's commitment to environmental concerns can save resources and can have a substantial impact on the environment. Chapter 17 discusses how environmental management systems have caused innovations that have improved environmental performance.

POLITICAL BENEFITS

The current permitting and reporting processes required by environmental regulations can be time-consuming and difficult. An ISO 14001 environmental management system provides a means for consolidating and normalizing environmental documents and data, thus allowing faster permitting and reporting processes to be implemented. The rationalization of processes will usually be accompanied by an increase in data accuracy and lend greater integrity to the permits and reports submitted to regulatory authorities. In recognition of this integrity, regulators will be much less likely to seek enforcement actions and will increasingly rely on a business's internal management should a violation occur.

A key area of political benefit arises from the self-assessment and self-management of the technical issues associated with the environmental aspects of operations. Frequently, a business expends resources to comply with environmental regulations, but no significant environmental benefit is realized by that business from those expenditures. In some cases, the technical approaches fostered by the regulations cause the business to focus on areas of activities that have minimal environmental impact, while other areas with severe impacts are left relatively uncontrolled. The EPA has recently recognized this dilemma by stating that compliance with regulations and improved environmental performance are not necessarily the same thing. One-size-fits-all regulations that prescribe what to do and how to do it can impede technical improvements that would otherwise lead to improving the environment.

Through both the enhanced communication and increased reporting integrity fostered by an ISO 14001 environmental management system, a business can address the technical management of processes unique to its industry and work with regulators to prevent ineffective and costly command-and-control initiatives from becoming regulations. The opportunity lies in moving the regulators away from a micromanagement approach, which is best controlled by the people closest to the processes, to a macromanagement approach. Industry needs to be allowed to seek its own solutions while continuing to meet specific performance levels required by law.

An ISO 14001 environmental management system can provide a common, systematic approach for international businesses working in countries with different regulatory frameworks. It is a practical means to proactively manage regulatory compliance, regardless of the content of those regulations. By adopting a simple, pragmatic method of accessing regulatory requirements and ensuring compliance with these requirements, the ISO 14001 framework encourages effective environmental management within a wide range of regulations and laws.

Within the U.S. a number of voluntary partnership initiatives between private industry and the EPA are in progress. Those partnerships, including Project XL, the Common Sense Initiative, and the Merit Partnership for Pollution Prevention, are exploring alternate regulatory approaches based on the use of voluntary consensus standards, such as ISO 14001. These initiatives and others are discussed more fully in Chapter 14. Although the EPA is struggling with its conflicting roles of enforcement and technical assistance, implementation of ISO 14001 industry-wide could reframe the future regulatory context by including voluntary consensus standards as an integral part of the EPA's programs. The revised EPA audit and self-policing policy, discussed in Chapter 11, is an important step in recognizing the value of providing incentives to businesses to be more proactive as stewards, not just users, of the environment.

TECHNOLOGICAL BENEFITS

As a management system, ISO 14001 can be considered a technology in itself. Although the focus of the standard is management systems, these systems are intended to improve environmental performance by preventing pollution and removing systemic causes of noncompliance. The systems approach looks at the interrelationships of many activities, and how they work together to achieve implementation of the environmental policy. By analyzing processes for their environmental impacts, management obtains a clear picture of how the business functions with respect to the environment. Management can then make better decisions about allocating scarce resources to minimize or eliminate negative impacts.

When management focuses on aligning processes with stakeholder requirements for environmental performance, it reveals both performance gaps and improvement opportunities. How a business can conduct a gaps analysis is considered in Chapter 18. Decisions about environmental technologies should be internalized, so that management can easily exceed the regulatory requirements and integrate environmental issues into business decisions. Both of these elements are the foundation for continual improvement of environmental performance, and foster the integration of technical environmental activities and overall business strategies.

A central feature of an ISO 14001 environmental management system is that it provides a single environmental management system for global organizations, or businesses with multiple sites and facilities. Consider the example of financial management systems. How would most businesses function if they paid their salaries and expenses at different times using different forms of currency? What if they reported their activities irregularly, incompletely, and not according to a standard format? Most businesses would disintegrate into chaos with such a fragmented financial management system. By using a standard environmental management system, the business can define and control its environmental impacts, and track them in ways familiar to financial managers.

IDEOLOGICAL BENEFITS

Ideology is characterized by the ideas and concepts that societies throughout the world embrace as interpretations of reality. These ideas include religion, science, philosophy, and the arts. Emerging ideas and concepts in these areas have foretold massive shifts in understanding such as the Renaissance, the Industrial Revolution, and now the Knowledge Revolution.

Beginning with the first Earth Day in 1970, and more recently the Rio Conference in 1992, an emerging awareness of environmental responsibility has been taking hold in the industrialized and developing nations of the world. This multi-faceted idea consists of several concepts: sustainable development, ecological integration, intergenerational responsibility, and natural resource stewardship. People throughout the world are searching for solutions to environmental problems involving the contamination of air, water, and soils. They are uniformly rejecting the throw-away, consumer mentality of the 20th Century. The synchronism of their ideas demonstrates that universal environmental reform is what people want, and that we can use the same solutions to ecological problems thoughout the world.

ISO 14000 represents a series of methods to transfer these ideas and concepts into the business context. The usable management systems of ISO 14000 can be implemented into operations and provide a bridge between the heavily polluting, resource-intensive industries that characterized the early phases of the Industrial Revolution and the new, cleaner technologies of the 21st Century. Businesses that understand and embrace the concepts of sustainable development, ecological integration, intergenerational responsibility, and stewardship will find ISO 14001 to be an indispensable means to achieving these important goals. Wealth in the future will be created from managed information, not from depleting natural resources or manufacturing (Mcinerney and White, 1995).

FINANCIAL BENEFITS

In an ideal world, all people and businesses would manage their environmental affairs purely for moral, ethical, and social reasons. The public is now beginning to realize, however, that the best way to achieve environmental change is to reward effective, proactive environmental management. Positive incentives, as well as clear direction, are crucial in the attainment of environmental goals. Financial reward is an acceptable and appropriate motive for environmental improvement.

Financial benefits of effective environmental management can be attained by increasing consumer and shareholder confidence; reducing the costs of doing business; improving relationships with investment bankers, commercial lenders, and the stock and bond brokerage community; boosting management and employee morale; increasing profits; and cutting legal and administrative costs. These benefits can have a substantial and long-lasting effect on the financial viability of a business.

The ISO 14001 standard for environmental management provides a practical and workable framework for controlling environmental risk. Its focus on continual improvement and pollution prevention encourages businesses to move from reactive risk management and risk financing into comprehensive risk-control activities. Following this path, businesses assure themselves and stakeholders, including financial partners, that they are identifying, prioritizing, and actively managing environmental exposures to lessen the likelihood of loss. Financial stability is an important byproduct of a well-managed environmental system.

COSTS OF IMPLEMENTATION

As with any investment, management needs to be aware of the operational and financial consequences that the development and implementation of an ISO 14001 environmental management system will have on the business. One of the benefits of implementation can be the identification of costs associated with environmental activities, thus enabling management to make more logical decisions about pollution prevention and waste minimization alternatives.

The first major category of implementation costs is internal resources. These costs usually represent the majority of the expenditures, and include personnel time, training, and information technology to support the new flow of environmental information. Personnel will be involved in developing documentation, defining process and information needs, and managing the project and its myriad of activities. Depending upon the results of the initial review and process analysis, more measuring and monitoring equipment or upgrades to current equipment may be required. Equipment costs can be capitalized and amortized, while personnel time is generally an operating expense in the period during which it occurred. Internal resources are approximately 80% of the costs of implementing an environmental management system.

Although the majority of costs usually relate to internal resources, many businesses find that they do not possess the required expertise to fully develop the required systems. A business may use external consultants and purchase commercially available training programs for its employees. A business may find that investments in outside expertise can reduce implementation time considerably by providing critical assistance to the staff who will be tasked with developing and implementing the environmental management system.

Should a business seek registration under the ISO 14001 standard, it will require the services of a registrar. These services are both one-time and continuing. One-time costs include the initial pre-assessment and registration audits. The business also needs to demonstrate continual conformance with the standard, since full recertification is required every three years under the current accreditation scheme. Management can also plan on periodic (every six months) conformance audits performed by the registrar as a cost of maintaining registration.

Virtually every business faces the possibility of environmental liability costs. Costs may be derived from lawsuits involving customers, employees, or communities or from legally mandated clean-up of hazardous waste sites. It is essential that senior managers make at least a general estimate of their business's potential future environmental liability. The U.S. Securities and Exchange Commission (SEC) requires adequate disclosure of environmental matters with the Commission (Armao and Griffith, 1997). A logical and accurate assessment of potential liabilities will not only allow for the allocation of suitable levels of financial resources, but will also articulate a comprehensive risk management program and a reassessment of corporate strategy and management practices. How to manage costly environmentally contaminated sites using ISO 14001 to improve decision making is the subject of Chapter 9 and avoiding costly litigation is the subject of Chapter 12.

The initial development of an environmental management system is time-consuming and expensive, but costs a fraction of potential litigation and clean-up costs. If an environmental management system is properly implemented, the investment will result in impressive annual savings derived from improved management efficiencies.

From a financial perspective, a business should determine how its environmental performance is viewed by shareholders, lenders, insurers, and employees. Financial risks include loss of profit, loss of access to capital, loss of market share, and loss of business. Benefits include an enhanced ability to survive, succeed, and prosper.

Businesses that have achieved ISO 14001 certification will have demonstrated a positive commitment to managing the immediate and long-term impacts of their products, services, and processes on the environment. ISO 14001 conformance should favorably impact the cost and availability of environmental liability insurance products. Businesses that demonstrate through ISO 14001 conformance that they are environmentally proactive should receive favorable underwriting consideration and should qualify for a decrease in premium levels. Several environmental insurance companies have already committed to include ISO 14001 certification as a critical factor in their environmental risk management underwriting processes.

By implementing an environmental management system, senior management ensures that it is effectively handling its environmental risks and limiting its exposures, addressing structural solutions, and reducing the need for excessive risk financing. Environmental loss controls, such as those promulgated by the ISO 14001 management systems, place an entity in the best position to access environmental insurance or other risk-financing vehicles for an acceptable cost.

Cost/Benefit Analysis

As part of the process analysis, management must analyze various environmental costs, including hidden expenses, in reaching better decisions regarding systems operations. The key to evaluating the investment in an ISO 14001 environmental management system is to derive a set of costs that accurately reflect current processes, and match them with a set of benefits incurred as a result of the environmental management system. In order to do this, management should attempt to match the temporal, quantitative, and qualitative characteristics of costs and benefits to more clearly understand their relationships.

The full range of environmental costs for an environmental management system will include both implementation and operating costs. Once the initial expenditure for system development is made, ongoing costs would include the following:

- Payroll and direct personnel expenses associated with system activities
- Environmental management system and compliance audits
- Environmental testing
- Purchasing, calibrating, and operating environmental monitoring equipment
- Environmental clean-up costs
- Environmental liability insurance premiums
- Waste disposal costs
- Reserves set aside for contingent liabilities

The key to performing an effective cost-benefit analysis is the ability to calculate the specific environmentally related costs associated with producing a particular product or service. Analyzing the costs and the benefits realized from implementing the environmental policy and strategy can answer two basic business questions for management:

- Have the environmental activities resulted in an increase in sales?
- Have the environmental activities resulted in a decrease in operating costs?

The environmental management system must demonstrate that it contributes positively to the operations and financial status of the business to be accepted as a reasonable investment. A typical presentation to management will show one-time investments resulting in long-term lower operating costs. Standard financial evaluation techniques, such as the Internal Rate of Return or Return on Investment, can be used. Indirect links to increased sales could also be shown to further favor the investment.

The problem with evaluating environmental management system benefits is that even though the costs incurred will be substantial and immediate, many of the benefits will be long-term and not completely quantifiable. Many direct cost reductions in waste disposal and energy and material usage could be realized, while other environmental management system activities could be placed under the rubric of risk prevention. In evaluating the financial performance of risk-prevention activities, management will have to look at the opportunity costs or avoided costs of preventive activities in their analysis. These opportunities and avoided costs can be difficult to estimate, but should not deter management from making a basic evaluation of their magnitude.

Many businesses have redesigned their processes to eliminate toxic chemical use, thus avoiding the direct costs of disposal and pollution control equipment altogether. More difficult to estimate, however, is the amount of fines, penalties, legal fees, clean-up expenses, and revenue loss because of negative public relations that the business would have incurred had it continued to operate by utilizing the old procedures. In this type of analysis, it is important to remember the beneficial stakeholder relationships a business can gain from communicating the results of its environmental activities, including customer acceptance and loyalty — and maintaining a favorable corporate image.

Domestic and International Industry-Specific Certification Advantages

Certification can benefit specific industry sectors, both in the U.S. and internationally. Each industry sector is different and faces unique environmental challenges. Some of the industry groups that are most impacted are General Manufacturing, Chemical Manufacturing, Electronic Equipment and Component Manufacturing, Service Industries, and Health and Pharmaceuticals.

General Manufacturing

Within the U.S. approximately 9000 businesses are currently certified to one of the ISO 9000 series of quality management standards. The primary incentive for ISO 14001 certification for these businesses, which make everything from doorknobs to automobiles, is customer pressure. Although manufacturing firms have already realized process efficiency gains as a result of implementing an ISO 9000 quality management system, they are often hesitant to embark on the 14001 journey unless their customers pressure them to do so. Maintaining regulatory compliance is generally their main goal, as well as and an acceptable strategic position. ISO 14001 certification would primarily benefit this industry group by allowing them the competitive advantage of remaining on the acceptable supplier listings.

Chemical Manufacturing

Chemical manufacturers have long been aware of the high risks posed by their industry, and were among the first to adopt various types of environmental management systems. This group has also been one of the most heavily regulated, resulting in a tangle of activities that are sometimes at cross-purposes with each other. While this group may also feel customer pressure to provide ISO 14001 certification as a condition of being an acceptable supplier, ISO 14001 certification could significantly streamline, integrate, and improve their existing quality and environmental management systems, resulting in lower environmental risks and costs. This integrative activity would eliminate redundancies and inefficiencies arising from a regulatory compliance focus, allowing many firms to address their environmental issues in a more comprehensive manner. Insurers may also pressure toxic chemical users and manufacturers to certify in order to demonstrate responsible risk management.

Electronic Equipment and Component Manufacturing

Electronic equipment and component manufacturers may also feel customer pressure to demonstrate ISO 14001 certification as a condition of doing business. However, the emerging benefit for these types of firms is obtaining a competitive advantage through marketplace positioning. Green products and businesses are becoming increasingly prevalent in this industry as businesses seek additional ways to differentiate their product offerings. In some instances, the issues these businesses face are similar to those of the chemical industry. They could realize many of the same benefits by improving their environmental management systems and streamlining operations.

Service Industries

Service industries face completely different environmental management issues. For these businesses, market positioning is the central environmental issue and a source of competitive advantage. Customer perceptions form the backbone of the purchase decisions making these businesses viable. A certified environmental management system will allow service industries to clearly demonstrate their proactivity and responsibility with regard to environmental management, as well as advertise the international recognition that comes from obtaining registration. Service industries may also benefit from the improved public relations that an environmental management system can generate. How they are perceived by their customers in the local economy can have major trickle-down effects on other customers located in geographically dispersed areas.

Health and Pharmaceuticals

The central issue for most health and pharmaceutical firms has always been quality. An environmental management system can provide integration of related, but not always considered, issues into the existing management structure. While the key benefit to these businesses would be to maintain status as an acceptable supplier, another benefit would be in resource productivity. Integrating quality and environmental management systems provides a comprehensive focus on increasing the efficiency of the production processes and reducing waste and pollution. Process improvement becomes a major issue, with yields expressed in concrete dollar amounts — especially for businesses with multiple manufacturing sites in different states and countries.

International Advantages

When considered from an international perspective, the advantages of ISO 14001 certification change considerably. Within the industrialized nations, the overriding benefit becomes the ability to reduce pollution and maintain social credibility. The already developed structures for certification provide an important recognition status for ISO 14001 implementation. When considered from the macro perspective, however, the main benefit to an ISO 14001 environmental management system is the demonstration of responsible environmental management. Management can also be assured through process efficiencies, reduction of waste, and proactive communications that the business is perceived as a good neighbor and a preferred supplier by its customers, communities, and markets.

In the developing countries, ISO 14001 certification is considered an asset to international competitiveness. Businesses located in these countries need to secure a leading-edge position with regard to modern technology. Obtaining ISO 14001 registration encourages advancement in the world marketplace and removes stigmas associated with outdated environmental management practices. Export markets are crucial to the success of emerging nations, especially for the extractive and manufacturing industries. These businesses can use the management technology presented in the ISO 14001 standard to advance not only their environmental practices, but also their overall strategic and operational management planning.

As the ISO 14001 environmental management system standard replaces the plethora of burgeoning national and regional environmental management standards worldwide, developing countries' industries will realize a distinct trade advantage by obtaining certification. The fewer standards with which a business must comply, the easier it will be to sell its products throughout the world. As the developing countries' industries grow, they also have an opportunity to build in their environmental management systems from the very beginning, thus obtaining a cost advantage that increases with time. Adoption of the ISO 14001 standard by developing countries can ensure that the push for improved corporate environmental quality does not become a hindrance to international trade.

The benefits of developing a system to manage and reduce environmental risks will inevitably outweigh the costs. The advantages can include increased productivity and morale as a result of joint enterprises that positively impact the environment. As discussed in the next chapter, these positive impacts can also include the development of a single integrated strategic information system that can be understood and fully utilized by management, employees, and customers.

Chapter 6

STRATEGIC INFORMATION FOR RISK MANAGEMENT SYSTEMS

Strategic information management is the integration of the complex array of environmental data existing within most businesses to support management decision making. Management needs accurate and complete environmental information to facilitate regulatory reporting and environmental communications, and to increase the effectiveness of decisions regarding environmental performance. With information integration comes an increased awareness of environmental management activities—leading to the identification of opportunities to eliminate redundant or polluting operations, to minimize energy and natural resource usage, and to address high-risk issues. Environmental risks are identified in this process and managed on the basis of credible data rather than speculation.

EVALUATION AND SELECTION OF INFORMATION TECHNOLOGY

Automated information technology supports an environmental management system and makes it possible to bring together the types of information necessary for effective environmental risk control. Even though the ISO 14001 standard does not specifically require information technology, it is difficult to imagine a business addressing its environmental activities without some form of automation. What the standard does require is a clearly stated and comprehensive description of how the business manages its environmental performance. Information technology makes such a description possible, and assists in managing the risks associated with the daily activities of all employees. Once environmental information is collected and analyzed using a database, then steps can be taken to reduce risk, as demonstrated in Chapters 10-17.

Before considering the specific information technology to use, companies should first evaluate the following overall performance criteria required by an environmental management system. The technology must:

- Provide complete information about the environmental performance of the business
- Meet all of the current regulatory and organizational reporting needs
- Establish links among the different environmental management system elements

- Provide access to functional information about the elements of the environmental management system by a wide variety of personnel
- Support the decision requirements of personnel responsible for environmental management, as defined and documented in the environmental management system
- Support the transaction needs of each environmental management system element

Management must also consider three basic technological strategies to support its environmental management system:

- Modification and enhancement of existing systems
- Purchase and installation of a commercial system
- Development and implementation of a data warehouse-based system

Many businesses already use some form of environmental management information system. These systems, however, tend to be fragmented and optimized for particular functions, such as for the federal *Resource Conservation Recovery Act* (RCRA), hazardous waste management, purchasing, or regulatory reporting. Numerous technologies also purport to provide integrated information support for an environmental management system, but can often cause more problems than they resolve. Too much information in a decentralized context, or seemingly inconsistent data packages, can confuse decision makers and cause conflicts and disruptions in business operations.

Should management build an entirely new system based on the new environmental management system requirements? While there is no simple answer, the existing information systems should be evaluated against the six performance criteria listed above. At present, no acceptable standardized information technology solutions are available for overall environmental management as an environmental management system by its very nature must be customized to the activities and processes of the business.

Modifying or enhancing existing systems may be appropriate for recently installed or upgraded systems, but the integration requirements of the environmental management system may exceed the capacity of those systems or require reduced performance of individual functions. Businesses need to keep in mind that commercially available environmental management system packages need to be connected to the existing systems. A canned environmental management system package is not appropriate for mid- to large-size businesses with complex environmental regulatory obligations.

Many large businesses are implementing an enterprise-wide solution. The business creates a series of transaction-based modules that feed data into a central database or series of databases. The management can then access the databases by using a high-level query tool with the capacity to deal with complex or technically challenging issues. This type of solution can support records management, communications, and data trending and analysis for continual improvement of environmental performance.

PROCESS REENGINEERING

Any information technology development project should involve some form of process reengineering if it is to provide value to the business. An ISO 14001 environmental management system must integrate the environmental function into all relevant business processes so that they can be managed within the context of the strategic positioning of the firm.

During the initial analysis, redundant or inefficient activities in the process should be reduced, often with the help of an information technology systems analyst. It is much easier to support the operation of a rationalized process than to attempt to retrofit a solution onto a flawed procedure. Thus, management can assure itself that it has measured its processes against environmental criteria, and has built-in information technology solutions to integrate those processes on an organization-wide basis.

Other significant process management issues must be considered. Increased availability of information will change the way decisions are made about environmental performance, and can have tremendous collateral consequences as job functions change and as employees seek new ways to exercise decision-making power. Those who do not innovate with increased information (see Chapter 17) and who continue to use old decision processes may not stand up to the scrutiny of the systems analysis efforts. Excuses like "that's the way we've always done it" will become unacceptable.

The new set of decision processes will be made by employees who have access to more information about how the business works than ever before. These decision makers may have to work with strangers and make decisions seriously affecting other departments and functions. If the business is not already integrated and cross-functional, the information technology solution supporting the environmental management system will force these issues and may have tremendous cultural and political implications for overall organizational performance. Managers who have built empires around the control of proprietary information will find them crumbling as they are exposed to the web of highly integrated information.

A central issue for larger businesses is integrating data across corporate levels and within each business unit. This will reveal duplication of effort and redundant operations. Data integration requires, however, semantic consistency for all business units. This means standardizing the way data are defined, gathered, processed, and reported in order to establish a useful basis of comparison between operations or sites. Semantic consistency is also critical during the data compilation and reporting stages in order to ensure that meaningful summaries are prepared for senior management and to ensure that errors and dysfunctions are identified and corrected quickly.

Another process management issue is the need for training and awareness regarding how the business manages its environmental impacts. In order to function within an integrated environmental management system, the employees need to be personally aware of their individual impacts on the environment. This includes how they travel to and from the workplace; their personal energy consumption; and how they integrate recycling, energy minimization, and other waste reduction technologies into their daily business lives. Information obtained concerning their environmental performance, both individually and collectively, can improve the decisions they make and impact the overall functioning of the business.

Employees also need to structure decisions to achieve maximum environmental performance. Thus, training takes on several new dimensions beyond specific job functions, the most important of which is the management of innovations and change. Employees need to comprehend how management has chosen to structure the environmental management practices of the business, as well as their roles and responsibilities for continual improvement. Employees also need to know how their jobs are affected by environmental practices, what they are expected to do, and the reasonable expectations for organizational environmental performance. Employees' job performances should be analyzed on a yearly basis, and job evaluation criteria should include how they have been able to make progress in reducing their environmental impacts at the workplace. Companies can consider bonuses and other incentives for employees who substantially reduce their individual impacts. All of this information can be made a part of a central data system.

CONSIDERATION OF THE ELEMENTS OF THE ENVIRONMENTAL MANAGEMENT SYSTEM

Regardless of the information technology system selected, each element of the ISO 14001 standard also has specific requirements for the creation, dissemination, and retention of environmental information. The significance of each element and its attendant performance characteristics are detailed below.

Initial Review

Information should be gathered and analyzed from a wide variety of sources that can include inventory control, purchasing, chemical tracking, emissions reporting, permitting, natural resource and energy usage, transportation, and waste management and disposal functions. Businesses that have a query capability that can access the data from these separate processes can perform the initial review more easily than others. Regardless of the sophistication of an existing system, data collection is essential to understanding the scope of a business's environmental impacts. A realistic assessment can then be made of the cost and timing necessary to turn whatever system the company is currently using into an effective environmental management system that identifies and manages environmental risk.

Environmental Policy

An environmental policy is an excellent vehicle to set forth, for internal and external review, a company's capabilities to collect information on its own environmental impacts. In creating a sound policy, management teams, in conjunction with employees, need to develop an information technology strategy that provides for the dissemination of policy information to all levels of employees, the public, and other stakeholders. The policy can identify various communications devices, like e-mail or file-sharing communications functions, to assist this effort.

Environmental Planning

The information technology strategy should support the business's environmental aspects and provide a means of describing processes and their relationships. Especially important is the inclusion of a life-cycle modeling capability to perform sensitivity analyses to determine where and how environmental impacts can be minimized or eliminated.

The information technology strategy can support the tracking and evaluation of legal requirements, including a method to evaluate the demands of changing regulations. The strategy also needs to incorporate organizational requirements that impact environmental performance, such as community commitments, agreements, and consent decree actions.

The automated information technology can greatly assist management with the control of objectives, targets, and programs. The strategy should be able to identify:

- Current goals, objectives, standards, and trends of industry
- Progress against those goals, objectives, and industry averages (benchmarking)
- Priority activities and their relationships to other areas of organizational activity (marketing, sales, finance, or operations)

Environmental Operations

The information technology strategy should provide an interface to the environmental training function and report on the current status of each employee's environmental training and job performance. Staffing requirements, scheduling, and selection need to be made available, as well as a means for evaluating skills and placing personnel in appropriate positions.

The information technology strategy should also facilitate environmental communications with desktop publishing, World Wide Web access, netware such as Lotus Notes®, and information-sharing bulletin boards. The Internet can also ease external communications. Companies can consider placing the environmental policy on a web page and summarize important parts of their environmental management systems for access by the public, other businesses, suppliers, and vendors.

Emergency Planning and Response

The information technology strategy should support the documentation requirements of the environmental management system, including regulatory reporting, recording of operating conditions and environmental exposures to employees, and both overall and process-specific emissions levels. A key consideration is the integration of all documentation activities related to environmental performance, including such areas as market research and customer requirements, product and service design, subcontractor selection and evaluation, manufacturing process development and control, and product use and disposal operations.

The information technology strategy should support emergency preparedness and response activities, including sensors to detect excessive toxic releases. The management of emergency response activities can be assisted by site mapping, personnel identification, hazard determination, and external communication functions.

Monitoring and Measuring

Integration of automated data collection from process monitoring equipment, such as sensors and detectors, is crucial to obtain a clear picture of environmental impacts such as emissions. The information technology strategy should also include functions for tracking and trending results of compliance and environmental management system audits, as well as the corrective actions resulting from these audits.

Management Reviews

The information technology solution should assist management by gathering and reporting all of the data generated by the environmental management system into some form of executive information system for review. The necessary capability will bring together diverse sets of data to support strategic decisions regarding environmental performance and the management of attendant risks.

The information technology strategy is an important component of the environmental management system. Information that is mishandled or miscommunicated can be devastating to system operations. Companies need to design effective risk communications components to complement the information technology system, as discussed in the following chapter.

Chapter 7

ENVIRONMENTAL RISK COMMUNICATION

One of the most important components of any environmental risk management strategy is the definition and control of organizational communications methods. Regardless of the actual risks posed by a business's activities, it is the stakeholder perceptions of these risks that can mean the success or failure of the environmental management system. These risk perceptions, which drive community and public interest group activities, can only be managed effectively by a clearly defined set of communications principles, procedures, and goals. Thus, management must develop and implement a comprehensive communications strategy aligned with the business's environmental policy, impacts, and regulatory requirements.

Effective environmental communications strategies have two primary benefits. First, they provide the opportunity to reduce duplicative data collection, reporting, and recordkeeping. Effective communications also benefit the overall strategic positioning of the firm. By evaluating and rationalizing its information needs, the business establishes the basis for implementing an automated information management technology (discussed in Chapter 6). An environmental communications strategy can provide much needed synergy between an information technology initiative and environmental management activities.

DEFINING THE BOUNDARIES OF THE COMMUNICATIONS STRATEGY

Businesses can select two basic forms of communications. Participatory communications require the active involvement of the audience, and establish a two-way system that allows the sharing of ideas. Interactive communications need facilitation and active solicitation of audience feedback to ensure that a common understanding is developed and comments are given adequate consideration.

The second form of communication, nonparticipatory, is a one-way message delivery to an intended audience. It is generally limited to providing information while the audience remains passive. Nonparticipatory communications do not ensure that a common understanding is achieved, but may be appropriate for many situations in which the business is responding to requests for information or is providing routine information required by regulatory agencies.

The ISO 14001 standard requires two methods of communications: internal to the business (management and employees) and external to the stakeholders. Both forms of communication should be founded on a common set of facts and ideas to ensure consistency, accuracy, and completeness. A functioning environmental management system can provide the data upon which the internal and external methods are built. Internal and external communications involve three basic types of environmental performance information:

- Internal environmental performance measures as part of the environmental management system
- Regulatory compliance and reporting
- Corporate and business unit environmental reporting requirements

For both internal and external communications programs, the business needs to define:

- Who is responsible for collecting the data
- How and where the data will be collected
- The specific data elements and types to be collected
- Who needs to receive the data
- What types of decisions will be made using the data
- Who are the intended audiences for the data

Once these program elements have been defined, the business can develop and implement specific communication methods based on their effectiveness with intended audiences.

ENVIRONMENTAL LABELING

There are approximately 25 countries that have some form of environmental labeling system for products. According to a recent report, the labels tend to provide the "barest of information" on the environmental impact of the product and the information provided is often of "dubious validity" (Morris and Scarlett, 1996). Eco-seals can result in a *de facto* means of erecting trade barriers and can become a means by which individual companies or nations, or both, advance their own products at the expense of others. The report rejected the concept of an international labeling system that would be "no more reliable as a source of environmental information than any other eco-label. Indeed, it may even be less reliable, since locational differences in environmental impacts are likely to be more extreme" (Morris and Scarlett, 1996). For all these reasons, the content of any environmental claim or statement to the public regarding a product or service needs to be considered carefully before distribution.

The following guidelines from the ISO 14021 labeling standard provide criteria against which such communications should be assessed. The environmental claim should be:

- Accurate and nondeceptive
- Substantiated and verifiable
- Relevant to the particular product or service, and used only in an appropriate context or setting
- Specific and clear as to what particular environmental aspects the claim relates
- Unlikely to result in misinterpretation
- Significant in relation to the overall environmental impact of the product or service during its life cycle
- Presented in a manner that clearly indicates that the environmental claim and explanatory statement should be read together
- Presented in a manner which does not imply, unless justified, that the claim is endorsed or certified by an independent third-party organization

The 14021 standard also suggests that vague or nonspecific claims that imply a product or service is environmentally beneficial or benign should be avoided. Communications should limit the use of the words *environmentally safe, environmentally friendly, earth friendly, nonpolluting, green, dolphin friendly, nature's friend,* and *ozone friendly.*

The 14021 standard also guides businesses on how and when to make legitimate environmental claims. Management should specifically focus the environmental communication on the business's measurable environmental performance characteristics, as well as the following product and service topics:

- Recycled content
- Recycled material
- Reduced resource use
- Recovered energy
- Solid waste reduction
- Energy efficiency and conservation
- Water efficiency and conservation
- Extended product life
- Product reuse and refilling
- Product recyclability
- Design for disassembly
- Compostability
- Biodegradability and photodegradability

INTERNAL RISK COMMUNICATIONS

Internal risk communications are directed toward employees and stockholders, and should be focused on achieving three objectives:

- Demonstrating management commitment to responsible environmental management
- Responding to questions and concerns about the business's environmental management activities
- Increasing awareness about the business's environmental policy, objectives, targets, and employee environmental responsibilities

Businesses should establish an internal environmental risk communications program based on the overall internal communications policy of the business, as well as the environmental policy. This communications policy can usually be found in an employee handbook or a compilation of company policies. The contents and activities of the internal communications program can address employee concerns and needs, the environmental aspects and impacts of the business, and the ongoing environmental activities.

Establishing a set of roles, responsibilities, and accountabilities for environmental performance provides the foundation for internal communication activities, and drives the type and frequency of individual communication events. Management can ascertain employee concerns and needs about environmental management by using a survey, the results of which can also be incorporated into the management review process of the environmental management system.

Management must also identify the scope of internal communications activities regarding significant environmental aspects of business operations. While there are regulatory requirements defining the external reporting requirements for an environmental incident, the business must define exactly what comprises an incident for internal communication purposes. While not every incident may need to be communicated to employees, the results of incident investigations can play an important part in corrective and preventive actions, as well as in the management review. It is important to remember that incidents may already be the subject of rumors and the informal communication network. Management should therefore take a proactive role to ensure that accurate and timely information is distributed to employees and other stakeholders.

Internal Communications Methods

Internal communications methods can vary widely according to the size, type, and environmental aspects and impacts of the business. Some common methods include the following:

- Daily, weekly, or monthly meetings
- Newsletters
- Pay envelope messages
- Notices in the workplace
- World Wide Web sites
- Video conferences
- Training sessions, on-site and off-site
- Performance reviews
- Intra-net applications (e-mail)

EXTERNAL RISK COMMUNICATIONS

External environmental risk communications involve the development and presentation of an accurate account of the effective management of the company's risks. Widely varying groups of stakeholders need to understand how environmental risk is managed within the company and not by or as a result of outside influences. Credibility, timeliness, appropriateness, and completeness become key factors in establishing and maintaining an effective program.

Interested third parties such as environmental, consumer, community, or other groups are continuing to hold businesses responsible for their environmental impacts and can demand improved environmental performance if they perceive environmental risks as unacceptable. The major strategic benefit from a carefully crafted external communications program can be the development of cooperative relationships which encourage stakeholder involvement and foster an ongoing and productive dialogue regarding environmental activities and impacts.

The ISO 14001 standard requires that businesses make a decision about any external communication initiatives not related to interested party inquiries, and record the results of this decision. Management must define the scope and nature of external environmental reporting as part of the environmental management system, melding the stakeholder awareness and regulatory requirements into an overall program based on the environmental performance reported by the environmental management system.

The requirements for external communication are among the weakest in the ISO 14001 standard. The required process is essentially passive and merely states that a business must receive, document, and respond to relevant external communications from interested parties regarding its environmental aspects and its management system. This presupposes that interested parties are already aware of the environmental activities of the business, which is not always the case. Proactive businesses will recognize this requirement for what it really is: the need to establish and maintain an environmental risk communications program with stakeholders.

Principles of External Environmental Risk Communication

Six basic principles of environmental risk communication should be applied to any external communication program developed as part of an ISO 14001 environmental management system:

Unfamiliar risks are less acceptable than familiar risks. Any external communications program should detail information on the pollution or wastestreams, including where they come from, how they are integrated into the production process, what are their environmental impacts, and what are the plans for continual improvement. Comparisons with existing processes used safely by other businesses and industry standards also help to build credibility and better relationships.

Involuntary risks are less acceptable than voluntary risks. The communications program manager must acknowledge the stakeholders' involvement in environmental decisions. Stakeholders will thus feel increased personal power and a part of the environmental decision-making process.

Undetectable risks are less acceptable than detectable risks. The communications manager must state clearly just what comprises both the proper and improper operation of the business's environmental impacts. A description of the monitoring and measuring activities will greatly reduce tensions with unfriendly stakeholder groups.

Risks perceived as unfair are less acceptable than risks perceived as fair. The key to fairness is to respond appropriately to stakeholders regarding their perception of risks associated with the facility's operation. Response can take a wide variety of forms, including direct monetary contributions for industrial accidents, community support programs, and involvement with community and charitable organizations.

Dramatic and memorable risks are unacceptable. People tend to judge an incident as more likely to occur if they can easily imagine it or recall a similar instance. Oil spills in various parts of the world have been vividly portrayed on television. These events are experienced personally by a great many people who will fault businesses for negligent or voluntary conduct that results in environmental calamities. Companies need to understand in the first instance the possible consequences of their operations in order to clearly state the protective measures that the company is prepared to take to prevent the occurrence of an accident.

Stakeholders are less interested in risk estimation than in risk reduction, and they are not interested in either one until their fears have been validated. Discussions about how to reduce risk tend to be more productive and demand a higher level of stakeholder involvement than those concerned with estimating the actual risk potential. Discussions should be focused on solutions rather than on theoretical possibilities, and on concrete action items rather than on placing blame. Careful discussion of the issues coupled with collaborative efforts to seek mutual goals is necessary to ensure the stakeholders do not have their fears validated.

The external communications strategy needs to consider the legitimate concerns of both environmental advocacy groups and consumers. These groups demand changes in environmental management practices that they perceive pose unacceptable risks. Management must carefully plan and pursue communications initiatives targeted at such advocacy groups to reduce the potential development of costly, adversarial relationships.

External Communications Methods

Methods for communicating with external stakeholder groups include the following:

- Public meetings with communities and chambers of commerce
- Annual environmental reports
- Community advisory panels
- Fact sheets and facility awareness information packets
- Press releases and articles in newspapers, magazines, and trade journals
- Media events
- Meetings with high-level public officials and regulators
- Customer service activities
- Facility tours
- Hotline numbers
- Trade association activities
- Internet applications such as World Wide Web sites

Each of these methods is self-explanatory, with the exception of environmental reports, which were mentioned in Chapter 2.

Environmental Reports

Environmental reports are an excellent way to transmit a company's environmental message and performance record to key stakeholders, including the employees and the general public. These reports may also be used to demonstrate publicly the continual improvement of environmental performance of the company. The following are summaries of five of the best current environmental reports:

ABB Environmental Management Report 1995 (Sweden)

ABB provides a description of its power generation, transmission, and distribution activities; its industrial and building systems; and its financial services in 43 countries. ABA has an advisory board that oversees all of ABB's environmental operations. Environmental management systems of 15 ABB businesses have been certified or verified, with 30 others in progress.

ASG 1995 (Sweden)

ASG is a transport and logistics business that describes its efforts to achieve *sustainable profitability* and *sustainable transport and logistics*. ASG profiles its strategic environmental alliances with Kodak and Baxter-Medical and benchmarks its performance in the U.S. and Germany. ASG also discusses its acquisition and integration of Frigoscandia into its environmental culture and systems.

Anglian Water (U.K.)

Anglian Water is the largest of 10 U.K.-registered water service businesses. Its report measures its performance for a number of years against a range of indicators and against industry norms.

GM (U.S.)
GM has produced two reports based upon the CERES reporting standard. (See Chapter 2.) It has a comprehensive benchmarking section, a section on environmental liabilities estimated at $223 million, and year-end reserves of $587.2 million for future liabilities.

Natwest Group (U.K.)
Natwest is a large bank with offices located worldwide. This report is particularly interesting because very few service businesses have prepared environmental reports. Natwest describes its comprehensive review of its efforts to reduce its environmental impacts, and generally, the challenge of environmental management for banking.

ENVIRONMENTAL COMMUNICATIONS STRATEGY

The development of the environmental communications strategy should follow the "Plan-Do-Check-Act" cycle of continual improvement. Each area must be defined and implemented to ensure comprehensive coverage of internal and external stakeholder environmental issues.

Plan
- Define the intent and purpose behind the program, including objectives and targets
- Define the scope, content, and type of the environmental communications (participatory or nonparticipatory)
- Identify the potential internal and external stakeholders and their interests and concerns
- Define a set of implementation activities necessary to match the communication of the business's environmental issues with each stakeholder's interests and agenda

Do
- Contact all stakeholders and obtain their involvement in and potential endorsement of the program
- Find out their specific agendas and interests
- Inform them about the business's environmental management activities
- Establish a format and mechanism for ongoing productive dialogues with stakeholders
- Establish a format and mechanism for community awareness and public and customer relations

- Establish involvement with ongoing community activities, such as civic groups, chambers of commerce, rotaries, nongovernmental organizations, and schools

Check

- Conduct periodic internal and external stakeholder surveys, community and customer focus groups, and public meetings
- Establish hotlines for environmental issues, and monitor stakeholder data and issues previously discussed
- Monitor media coverage of environmental issues related to the business
- Evaluate customer response to environmental strategies
- Review public statements or reports by interested parties

Act

- Conduct reviews of the external communications program as part of the management review process
- Implement changes as necessary for emerging issues or areas of concern and to ensure continual improvement

Well-developed information and communications strategies can play an important role in environmental management systems. Equally important is the ability of companies to eliminate liability risk altogether by using effective risk transfer strategies. The following chapter discusses how this is accomplished.

Chapter 8

INSURANCE AND RISK TRANSFER STRATEGIES

Businesses have various mechanisms to choose from to identify, assess, and manage environmental risk. Properly implemented environmental management systems can effectively accomplish all three of these business goals. Insurance is a critical component of any business operation. This chapter discusses how companies can manage environmental risk by transferring all or part of their environmental liabilities. Choosing the correct risk transfer strategy is a crucial business decision. Recently, many new options have emerged to transfer environmental risks to third parties.

Risk transfer can be closely related to a business's risk financing strategy. Each business must determine its tolerance for the self-assumption of risk and identify appropriate methods to distribute or transfer the remainder of the risk. A business choosing not to assume all of its own risks can transfer its current or potential environmental risks through a variety of mechanisms, including insurance, contractual risk transfer, and property transfer. These risk transfer mechanisms are not mutually exclusive, and the most cost-effective strategy is usually a combination of the various alternatives.

ENVIRONMENTAL INSURANCE

Insurance is by far the most common risk transfer mechanism. Billions of dollars in premiums are paid each year to transfer exposures to the insurance community. Businesses began purchasing general insurance in the late 19th Century, with each insurer issuing its own policy language. In the 1930s, the insurance industry began to standardize policy language, and in 1941, the first comprehensive general liability (CGL) policies were introduced (Bailey and Gulledge, 1997). For many years these policies were purchased by businesses and have resulted in insurers being held liable for environmental claims arising from disposal of wastes during this time period. In the early 1970s after Congress began enacting comprehensive environmental laws, the CGL policies were revised to exclude property damage from pollution or contamination that was not sudden and accidental.

The pollution exclusion provision contained in most general liability policies issued between 1973 and 1986 included the following language:

*This insurance does not apply to: bodily injury or property
damage arising out of the discharge, dispersal, release or escape of smoke, vapor, soot, fumes, acids, alkalis, toxic chemicals, liquids or gases, waste material or other irritants, contaminants…, but this exclusion does not apply if such discharge,
dispersal, release or escape is sudden and accidental.*

With the introduction of the absolute pollution exclusion and the total pollution exclusion in commercial general liability policies, it is now necessary to
add pollution coverage endorsements to these forms or to purchase separate environmental liability insurance if environmental risk is to be insured. Environmental impairment liability insurance came into existence in 1977 as separate
insurance coverage; however, the market for environmental insurance remained
relatively restricted until the late 1980s. With the opportunity to underwrite environmental risks, many insurers in the standard commercial property and casualty insurance market did not routinely add pollution coverage to general liability insurance policies, leaving the policyholder liable for environmental damages and restoration. If environmental coverage was provided by the general
insurance underwriter, it was commonly time element or sudden release coverage that did not include claims for damage to soil or groundwater resources.

Since 1986 the marketplace for specific types of environmental insurance
coverages has expanded rapidly. Currently, more than 25 varieties of environmental insurance are available, and more than 80 different environmental insurance policies are offered by pollution underwriters. Insurance products are available to cap remediation costs, to protect buyers and sellers from environmental
liabilities resulting from property transfers, to protect owners from known and
unknown environmental impairments, and to control insurance costs of environmental service providers. New forms cover claims arising from sudden and accidental, as well as gradual, releases. The following specific environmental insurance products are currently available:

- **Environmental Asset Liability (EAL)**: Coverage to protect property
  owners from the liabilities associated with known and unknown environmental conditions preexisting at a site.

- **Environmental Impairments Liability (EIL)**: Coverage to protect property owners from the liabilities associated with ongoing operations.

- **Remediation Stop Loss**: Coverage to cap remediation costs for known
  environmental conditions and limit risks to the owner. This cost cap coverage is designed to insure remediation costs that exceed the projected
  or anticipated costs in the execution of a remedial action plan at a specific location. This coverage is particularly useful in facilitating the sale
  of contaminated property.

- **Owner Control Insurance Program (OCIP)**: Coverage for construction or remediation activities. This coverage is held by the owner rather
  than the contractors in order to control costs.

Other types of coverage include:

- Contractors' pollution liability coverages
- Engineers' and consultants' professional errors and omissions liability coverages
- Asbestos and lead paint abatement contractors' insurance coverages
- Combined forms for professional and contracting risks

In addition to standard insurance coverages, environmental insurance policies are currently available to cover an extensive list of possible losses including:

- Emergency response actions
- Government actions
- Site assessment related to a pollution incident
- Remedial clean-up costs, including leaking underground storage tanks, PCB spills, and spills from stored hazardous materials
- Business interruption costs
- Monitoring costs
- Third-party litigation costs
- Remediation closure costs
- Change in state or federal laws
- Real estate stigma (diminution in value, guarantee of fair market value)
- Interim ownership securitization for the chain-of-title

The environmental insurance industry is in a state of transition. Any business that has not addressed its environmental insurance options within the past five years is probably paying too much for too little coverage. Premiums for specialty environmental insurance products have dropped between 20 and 30% in the last few years (Bailey and Gulledge, 1997). Due to the competitiveness of the environmental insurance marketplace, premiums have gone down and coverages have expanded dramatically. Although there are off-the-shelf insurance products available, the best way to insure risk is through a customized menu of insurance products designed to minimize a business's liabilities while not inflating premiums with unneeded coverages.

The implementation of an environmental management system such as ISO 14001 should favorably impact the cost and availability of environmental liability insurance products. Businesses that demonstrate environmental proactivity through ISO 14001 conformance should receive favorable underwriting consideration and qualify for a decrease in premium levels. The EPA's Merit Partnership for Pollution Prevention, which is more fully described in Chapter 14, is currently exploring opportunities for businesses to obtain new types of insurance coverage or reduced premiums on current coverages as a result of having implemented effective environmental management systems (Reich, 1997). The EPA is considering the potential impacts of ISO 14001 in providing better coverage at lesser rates. Several environmental insurance companies have already committed to include ISO 14001 certification as a critical factor in their environmental risk management underwriting processes.

Contractual Risk Transfer

A contract is an agreement between two or more parties supported by adequate consideration which creates a duty. The intent of a risk transfer contract is to have the risk reside with the party who has responsibility for action. However, the risk transfer mechanisms are not always put into writing prior to the loss, with the parties acting responsibly to resolve the loss according to the agreed-upon, loss-sharing arrangement. Often a judge or a jury allocates the loss. Unfortunately, when the effectiveness of a contractual risk transfer arrangement is determined by litigation, it can result in unpredictable and expensive outcomes.

Risk transfer allocates the loss potential of an activity or practice in a fair and equitable manner. The general theory is that "I will be responsible for my employees and property, and you will be responsible for your employees and property." Damage to third parties should be the responsibility of the person in control of the activity or operation. These simple concepts become complex when they are considered in the context of unequal bargaining power; federal, state, and local environmental laws and regulations; and changing market conditions. Contractual risk transfer techniques include hold harmless agreements, waivers of subrogation, exculpatory agreements, indemnity agreements, and contractual definition and division of liability agreements as outlined below.

Under a hold harmless agreement, a contracting party will assume the legal obligations of the other party from third-party actions. The hold harmless agreement becomes effective when one party (the indemnitor) agrees to hold the other party (the indemnitee) harmless from tort liability under third-party actions associated with the hazards of the venture. By virtue of the agreement, the indemnitor assumes the liability of the indemnitee. In a broad form indemnity, the indemnitor will be responsible for any and all liabilities, regardless of which party was at fault. In an intermediate form indemnity, the indemnitor assumes all liabilities with the exception of those due to the indemnitee's sole negligence. Under the limited form indemnity, the indemnitor owes the indemnitee for that portion of the claim related to his own liability. Although the limited form indemnity may make the most logical sense, it often leads to lawsuits where the courts apportion the claim since the percent or cause of liability will often be in dispute.

Waivers of subrogation are agreements between two parties that state that one or both parties' respective insurance companies have no right of recovery against the other party. Essentially, the insurance companies are made a part of the contractual agreement between the two parties. In this way, the parties can be certain that their insurance companies will not sue the responsible or negligent party, thereby circumventing the hold harmless agreement found in the contract. Generally speaking, whenever there is a hold harmless agreement in a contract, it should also be accompanied by a waiver of subrogation provision.

In an exculpatory agreement, one party accepts the entire blame caused by the other party and handles the entire loss or claim just as if the other party were nonexistent. Exculpatory agreements are common in equipment leases, mineral leases, land leases, right-of-way agreements, and water-use agreements.

An indemnity agreement protects against, and pays for, possible damage, legal suit, or bodily injury. Indemnities commonly involve the owner of personal property and the business that has temporary possession of the property. The holder of the property agrees to pay for any physical damage losses to the property owner stemming from the acts or omissions of the holder of the property.

The contractual definition and division of the liability risk transfer technique is an agreement specifying who will be financially responsibl. for certain losses. This type of agreement identifies potential losses and who will bear the financial responsibility for these losses.

Property Transfer

Many owners of contaminated property decide to divest themselves of environmental liabilities by selling their properties. These owners must realize that in the U.S. environmental laws impose joint and several liability for clean-up costs on all parties in the chain of title when there is a release. Therefore, selling a property may not reduce liabilities and may even increase exposure due to the likelihood of new discoveries. Sellers therefore desire to transfer properties to parties that have the resources to not only remediate the property, but also to protect the seller from possible future costs related to environmental liabilities of the property. For similar reasons, potential purchasers are extremely cautious about taking title to contaminated property.

When selling contaminated property, buyers and sellers often have difficulty in establishing the sales price, since there is commonly a wide discrepancy between the low and high estimates of clean-up costs. Whereas potential buyers tend to discount the sales price by the maximum potential remediation cost, the seller favors the low-cost estimate. Many real estate deals fail because of this lack of agreement. In this instance, remediation stop loss or cost cap insurance can be useful in facilitating the sale of the contaminated property. Remediation stop loss policies agree to pay, on behalf of the named insured, the expenses (in excess of the self-insured retention) that the insured incurs in completing an approved remedial activity at a specified location. Although the insurance has a price, using cost cap insurance can provide a remediation cost number that is guaranteed and therefore useful to both the buyer and seller.

The existence of risk management systems and new environmental insurance coverages has made possible a new industry of companies that buy contaminated real estate, known as brownfields, with an agreement to hold the seller harmless from all future environmental liability. The risk assessment, control, and management expertise of these risk assumption companies allows them to approach municipalities and regulators with an environmental action plan that focuses on reduction of environmental exposures. A municipality may be able to indemnify the new owner or occupant for historical contamination after remediation is completed, and insurance can supplement the indemnification, protecting the municipality against future claims associated with historical contamination. The lender providing remediation or purchase funding also can be protected against third-party claims (Bailey and Gulledge, 1997). This demonstrated ability to minimize exposure rationalizes the process and allows government officials to exercise whatever latitude they may have to expedite real estate transactions. Buyers of contaminated properties have multiple options to protect their interests, including remediation cap insurance covering the buyer for clean-up costs exceeding the site estimate; historical and ongoing coverage protecting the buyer from remediation costs associated with unknown and undiscovered contamination; and insurance for known releases and future site operations (Bailey and Gulledge, 1997). The next chapter discusses brownfields in greater detail to show how ISO 14001 can be used to improve the decision making, management, sale, and redevelopment of these contaminated properties.

Chapter 9

MANAGING BROWNFIELDS RISKS USING ISO 14000

Brownfields present complex questions of social, political, and economic policy for governments and businesses throughout the world. The expansion of insurance coverage for past and future environmental contamination, coupled with the change in the U.S. regulatory climate, has caused new policies to develop which are designed to provide incentives to businesses to redevelop contaminated properties. In this chapter we take a closer look at brownfields, the opportunities they present, and methods to effectively manage risk along the way.

THE BROWNFIELDS PROBLEM

Brownfields have been defined in proposed federal legislation as "parcels of land that contain or contained abandoned or under-used commercial or industrial facilities, the expansion or redevelopment of which is complicated by the presence or potential presence of hazardous substances, pollutants, or contaminants" (*Community Revitalization and Brownfield Cleanup Act of 1997*). Developers' past avoidance of brownfields because of uncertainties in the amount of clean up and development costs have contributed to construction on undeveloped greenfields sites, which create urban sprawl, infrastructure problems and reduce the amount of open spaces (*Community Revitalization and Brownfield Cleanup Act of 1997*). Brownfields exist in every country in the world and pose remarkably similar problems.

Some countries, like the U.S., are attempting to solve these problems by remediating existing brownfields and, at the same time, seeking to eliminate the causes of contamination in an effort to prevent or at least reduce the number of future brownfields. As mentioned in Chapter 3, the U.S. has been using command-and-control environmental laws and regulations for many years to force companies to clean up old contamination whether they were responsible for the contamination or not. Many businesses were reluctant to expend resources on old industrial sites and instead decided to move their companies to new locations, usually uncontaminated greenfields, or to simply shut down operations.

One exception is Howard M. Kilguss, the owner of Excell Manufacturing (a $40 million jewelry manufacturing company located in Providence, RI). Mr. Kilguss decided not to rebuild in a greenfields location outside the city, but instead relocated two of his jewelry factories to an industrial site within the city. Kilguss was concerned about the environmental implications of the move, which was made necessary because of the need to consolidate both factories. He was also worried about the impact on many of his employees who took public transportation to get to their workplaces in Providence. In relocating both factories to one industrialized site in the city, he measured the environmental, social, and political impacts of the move and decided that it was more important to reduce those adverse impacts than increase short-term profits. Kilguss went further and had his employees actually participate in the planning of the new, larger facility in an attempt to have all of the employees buy into the environmental considerations that underpinned the project. This employee support made the transition work better in the long term. Not all company owners have the same concerns for their employees or the environment.

It is estimated that about 10 large factories close each week nationwide, with an average of 190 workers losing their jobs with each closure. Federal, state, and local governments, with a great deal of input from business, have finally decided that innovations are necessary to solve such massive problems and that management systems, such as ISO 14000, are necessary to save and create new jobs in disadvantaged urban areas.

According to government studies, there are approximately 130,000 to 425,000 brownfields sites in the U.S. Philadelphia alone has between 1500 and 2500 properties that qualify as brownfields. On August 28, 1996 President Clinton announced, in the midst of his re-election campaign which emphasized his positive environmental record, a $2 billion plan to clean up brownfields. President Clinton has requested a brownfields tax incentive totaling $2 billion over the next seven years to fund the clean up of brownfields. The plan is to use these funds to give prospective purchasers new incentives to clean up and redevelop 17,000 or more brownfields sites in distressed communities across the U.S. by the year 2000 (Gore, 1997). Clinton stated that this project was "the most important thing I am working on with the mayors of America today." The President also promised to spend another $300 million on further study of brownfields and to get businesses financial incentives to clean up their brownfields.

According to Vice President Al Gore, the $300 million will be used to expand local site assessment efforts, support state voluntary clean-up programs, provide local training in remediation work, and expand guaranteed loan financing for brownfields sites through the Department of Housing and Urban Development (Gore, 1997). On May 13, 1997 the Administration announced the creation of the Brownfields Partnership of 15 federal agencies, which is expected to leverage from $5 to $28 billion in private investment, support up to 196,000 jobs, and protect up to 34,000 acres of undeveloped greenfields on the outskirts of cities throughout the country.

The vast number of brownfields in the U.S. requires new public-private sector partnerships to improve the quality of life and safety of the nation's citizens. The administration is seeking to promote such joint ventures by offering up to $2 billion in tax incentives for businesses to redevelop brownfields. The U.S.

Treasury Department has predicted that these incentives will revive 30,000 brownfields sites. There are some compelling economic reasons, aside from health and safety issues, why government is seeking to clean up these sites.

Many of these unremediated brownfields are in urban industrialized areas. There are obvious economic benefits to redeveloping these sites. For example, the redevelopment of brownfields in cities will create jobs and place them closer to dense population centers, thereby reducing urban unemployment and unnecessary transportation problems. "The mismatch between residential locations and the location of jobs is a problem for some workers in America because, unlike the system in Europe, public transportation is weak and expensive" (Wilson, 1996). Inner city redevelopment will increase municipal tax bases and lessen future suburban sprawl, according to the U.S. Conference of Mayors. That organization recently completed a 33-city survey which estimated that the cities cited lose potentially $386 million in taxes each year from unproductive industrial sites. A recent study of Chicago's brownfields found "Industrial redevelopment encourages environmental clean up, brings jobs to underemployed communities, recycles infrastructure, revitalizes deteriorating neighborhoods, and counteracts suburban sprawl" (Brownfields Forum, 1995). President Clinton correctly urged the nation to "clean up these urban toxic waste sites and turn them into homes for safe businesses that create jobs in areas that thought they would never get any new jobs again."

For years industry has been relocating to undeveloped rural areas, at times creating new polluted areas and requiring expensive extensions of infrastructure and municipal services into previously undeveloped land. In Chicago, for example, between 1970 and 1990 the regional population increased only 4% while the availability of urban land area expanded by 46%. The development of open space and farmland can have a negative impact on farming, natural habitat, air quality, energy consumption, and congestion. During that same period of time, total vehicle miles traveled in Chicago almost doubled because of sprawling land-use patterns. A costly public transportation infrastructure was required to accommodate the increased dependance on the use of automobiles. This particular scenario resulted in the Northeast Illinois area being in severe noncompliance with federal air quality standards. Six of the seven chief air pollutants come from automobiles according to the EPA (Colorado Commons, 1997).

The Chicago and Northern Illinois areas' problems of air quality based upon excessive automobile usage are consistent with national trends. The Federal Highway Administration has reported that from 1960 to 1990, U.S. auto travel increased 198% in miles traveled; there were 133% more registered cars; 126% more fuel was used; and licensed drivers increased by 91%, while the nation's population grew 39%. In 1960, 69.5% of Americans commuted by car. By 1990 that figure had risen to 86.5% and the use of public transportation decreased during that same period from 12.6% of all commuters to 5.3%. Even walking is decreasing—from 10.4% to 3.9% (Colorado Commons, 1997). The excessive dependency on the automobile is having these collateral consequences:

- Road congestion will triple in 15 years, according to the General Accounting Office — even if capacity increases only 20%.
- Driving delays are expected to waste 7.3 billion gallons of gas, increasing travelers' costs $41 billion and adding 73 million tons of carbon dioxide into the atmosphere.
- Pavement covers over 60,000 square miles in the U.S., or 2% of the total surface area.
- About 1.5 million acres of farmland is lost to suburban sprawl each year as a result of road building and car travel (Colorado Commons, 1997).

Federal, state, and local governments need to focus on transportation problems in conjunction with finding ways to redevelop brownfields so that jobs can be created closer to the labor market. New coalitions are now being formed to create programs and undertake other initiatives to find solutions to these environmental-labor problems. While current politically popular initiatives mostly focus on how to clean up existing brownfields, very little attention has been spent on developing systems to manage and prioritize the redevelopment of brownfields, to eliminate the causes of future brownfields, and to create opportunities for continuous improvement of the redeveloped land. A key component of any brownfields redevelopment management plan will be making certain that the businesses stay in their recycled homes so that properties have sufficient labor resources and unnecessary travel can be reduced.

BROWNFIELDS REMEDIATION

In February 1995 the EPA began its Brownfields Initiative, not by focusing on labor and transportation impacts, but on the stigma attached to these properties. The EPA deleted approximately 28,000 properties from its CERCLIS database, based upon its determination that these sites had a low likelihood of requiring further remedial action. The EPA determined that there was "no further remedial action planned" (NFRAP) at these sites under federal environmental law and acknowledged that maintaining the sites on the federal CERCLIS database had made it more difficult for lending institutions and real estate developers to redevelop these properties. The EPA decided "to counteract the market perception that, by being listed on CERCLIS, a site must have a significant hazardous waste problem and therefore should not be considered as a potential redevelopment site" (Abelson and McCaffrey, 1996).

Between 1995 and 1997, EPA has made grants to 113 communities across the nation to clean up the brownfields sites (Gore, 1997). The first project, in Cleveland, OH, created 200 new jobs and generated more than $1 million in new payroll taxes when a piece of old industrialized property was restored in that city. On November 1, 1996 the EPA began accepting proposals for the 1997 National Brownfields Economic Redevelopment Pilot projects. Each pilot project can receive up to $200,000 over 2 years to empower states, communities, and other parties interested in economic development to work together to prevent, safely clean up, and reuse brownfields.

Congress has bipartisan support for brownfields legislation that would fund clean up of brownfields throughout the U.S. The proposed *Community Revitalization and Brownfield Cleanup Act of 1997* provides a credit for clean up of certain polluted sites, and the use of tax-exempt redevelopment bonds to finance clean-up operations. Investors, businesses, and developers should take advantage of these opportunities because many have already recognized the benefits of redeveloping these sites. Some examples of successful redevelopment of brownfields in the city of Buffalo, NY include:

- American Axle Co. reclaimed a former General Motors plant and completed a $20 million addition on an adjacent industrial site.
- A closed GM plant was reopened by a number of smaller businesses.
- An $18 million retail complex was built on a former Atlas Steel plant site.
- A free-trade zone on a former plant site attracted 36 U.S. and Canadian industries.
- A former General Electric site will be used for a new inner city supermarket.
- A 30-acre hydroponic tomato farm is being built on an abandoned LTV steel mill (Tyson, 1996).

National policy initiatives to encourage brownfields development have been emerging. They include prospective purchaser agreements (PPAs). A PPA is an agreement where the EPA will agree not to sue (covenant not to sue) a prospective purchaser of a contaminated property if that individual or business organization agrees to clean up the site if the PPA results in a substantial direct benefit to the EPA. The EPA defines such ancillary benefits as the creation or retention of jobs, productive use of abandoned property, or revitalization of blighted areas. The most recent policy expands the original 1989 policy, which was underutilized and generally regarded as ineffective.

On July 3, 1995 the EPA issued its *Policy Toward Owners of Property Containing Contaminated Aquifers (Aquifer Policy)*. The EPA agreed not to take any enforcement action against a property owner if hazardous substances came to be located on the property solely as a result of subsurface migration in an aquifer from off-site sources. "EPA's intent was to issue a policy indicating that EPA would not expect to sue such a property owner or operator, with the intended result that the market for these properties would improve" (Abelson and McCaffrey, 1996).

These federal programs and policy initiatives have been supplemented by administrative actions on the state and local level. Some examples shown below will demonstrate a commitment by government at all levels to solve the brownfields problem.

State and Local Actions

Massachusetts is a good example of innovative state programming. In 1994 that state launched a pilot brownfields initiative in which government officials worked with representatives from business and the private sector to encourage the redevelopment of contaminated sites in designated economic target areas (Abelson and McCaffrey, 1996) . Massachusetts has an estimated 7000 brownfields sites. The intent of the initiative is to reduce the number of brownfields and to develop state policy to recycle these sites into productive properties. Similar voluntary clean-up programs have emerged in many other states including California, Illinois, Michigan, Missouri, Ohio, Pennsylvania, Colorado, Texas, and Wisconsin. Minnesota's program is the oldest, having recycled 1500 acres of industrialized brownfields since 1992. Under the *Pennsylvania Land Recycling Act of 1995*, 64 contaminated sites have been cleaned up and another 195 are in progress. By comparison, in the last 16 years under the Superfund program, only 8 of Pennsylvania's 103 sites have been cleaned up and removed from the Superfund National Priorities List.

In Massachusetts the applicant for the project must be a prospective purchaser or a tenant (not a "responsible party" with potential liability for the clean-up of the site) who is purchasing or leasing property in a disadvantaged economic target area. Massachusetts agreed not to sue the prospective purchaser or tenant if the property is cleaned up in accordance with Massachusetts law (the Massachusetts Contingency Plan or MCP) which addresses risk-based, clean-up standards. The process has no specific state oversight. Licensed Site Professionals (LSPs) from environmental consulting firms are retained to move the sites through the MCP process. This new program won an award in 1995 from the Council of State Governments and is now being followed by a number of other states including Connecticut, Illinois, and Ohio.

Cities like Chicago that have huge numbers of brownfields are engaged in local efforts to remediate these sites and to revitalize significantly deteriorating urban areas. In November 1993 the Chicago Department of Environment, Planning, and Development; the Law Department; the Department of Buildings; and the Mayor's Office formed an interdepartmental working group on brownfields (Brownfields Forum, 1995). The working group decided:

- To devise more responsive environmental and economic development policies
- To create a brownfields pilot program to clean up and redevelop demonstration sites in distressed neighborhoods
- To develop economic models that account more accurately for environmental and social cost benefits of development decisions (Brownfields Forum, 1995).

The working group found that the demand for industrial space in Chicago greatly exceeds its supply and that most of the hundreds of abandoned industrial properties within the metropolitan area are not competitive with greenfields sites in the suburbs and beyond. The working group further found that the fear of environmental costs and liabilities produced a stagnant marketplace where brownfields lay dormant for years. The working group chose five pilot projects to determine what steps would be necessary to recycle these properties into reuse. Of the five selected properties, one was found to be clean, while another had only minor contamination. When these and the other pilot properties were returned to productive use, the working group determined that these properties would create jobs that are especially valuable to the distressed communities where brownfields are commonly located. "Brownfields redevelopment can produce a halo effect, attracting additional investment in local businesses, public infrastructure and employment training" (Brownfields Forum, 1995).

The working group further found that communities needed to "close the loop" that links environmental remediation, redevelopment, and redevelopment activities to reap the full development of brownfields clean-ups. "If there is one over-arching theme to the Chicago experience, it is the need for a coordinated, comprehensive effort involving all key stakeholders. No one group can solve this problem alone. City, state and federal agencies have an integral role to play. So do banking, business, manufacturing, legal, insurance and real estate professionals; community industrial and economic development groups; trade associations, environmental and public interest groups; environmental justice representatives; organized labor; and community health organizations" (Brownfields Forum, 1995).

Following an extensive study, the Brownfields Forum came to these conclusions:

- Brownfields redevelopment should foster healthy communities throughout the city and region.
- Public incentives for greenfields development should not outweigh incentives for recycling brownfields. Redevelopment of brownfields areas will reduce the need for new infrastructure in outlying areas, conserve environmentally sensitive areas, and otherwise save the costs of sprawl.
- Engaging the private sector and expanding market resources are critical to brownfields redevelopment.
- Effective strategies require strong partnerships among government, communities, and the private sector.
- Public brownfields expenditures should:
    - Address sites that would not be redeveloped without government participation
    - Redevelop disadvantaged areas
    - Focus on areas where reuse of brownfields is likely to catalyze additional redevelopment
    - Create and retain jobs
    - Maximize public benefit
- Redevelopment efforts should seek to attract environmentally sound industries to prevent the spread of brownfields and to foster sustainable communities.

- Brownfields initiatives should be viewed as one important component of a comprehensive strategy for revitalizing urban communities.
- Brownfields redevelopment should seek to leverage broader, integrated strategies for providing viable, long-term and area-wide development.
- Environmental clean-up standards must be clarified to accommodate a full range of land-use options.
- A large-scale brownfields redevelopment program should be based on knowledge and experience gained through pilot efforts and tests of innovative approaches and tools.

The federal, state, and municipal experience shows that all three levels of government are seeking to provide substantial financial incentives that will result in the redevelopment of property and the creation of new jobs. One common theme among all these programs is that government officials have endeavored to study the brownfields problem in urban areas to understand what needs to be done to clean them up.

ISO 14000 AND BROWNFIELDS

The ISO 14000 standards may be used to facilitate this planning process, speed up the remediation of these sites, and manage the properties or recycle them for future use. ISO 14000 environmental management systems may assist these government entities and responsible businesses in addressing the brownfields problem on a massive scale and in using scarce resources to recycle these properties into productive use. Before money is wasted by poor planning and excessive transactional costs, it would be useful to have a policy that would allow for cost-effective and permanent solutions to brownfields problems.

Businesses that own or are responsible for brownfields can benefit by taking stock of the number of these sites and by devising the methods and means to manage these properties effectively until they are remediated on a prioritized basis, using risk-based corrective action, and put back into productive use.

No business with brownfields sites should embark on a brownfields redevelopment program without planning a comprehensive and consistent approach to manage the multiple risks associated with the problem. A business seeking to understand its brownfields problem, or a business intending to redevelop another company's brownfields properties, can use ISO 14000 to develop an environmental policy, objectives, and targets to effectively manage the risk involved.

An ISO-Brownfields redevelopment program would have as its goals increased productivity of property; creation of jobs for nearby residents; reduction of costs, wastes, and energy usage; prevention of pollution and unnecessary further industrial development; and assurance of responsible future use of the property. The goals must be clearly articulated in a policy by owners and managers who are seeking to develop appropriate standards to revitalize these distressed properties. Partnerships and joint ventures can be established to foster swifter and more cost-effective solutions to brownfields problems. ISO 14001 provides the foundation for these important business-social-economic decisions to be made.

The objectives of the program are (1) to reduce risk to abutters, neighbors, and employees if there are ongoing operations, using risk-based corrective action standards; and (2) to improve these properties for future productive use. Risk-based corrective action simply means that the clean-up standard is dependent on the future use of the property. More stringent standards are used for parks and residential areas as opposed to paved-over factory or commercial sites. According to the *Washington Post,* "[t]wo of the country's largest outlet shopping malls are being built on brownfields in Carson, CA and Elizabeth, NJ that no one would consider for homes." These projects, and others on a larger scale, need careful planning. ISO 14001 can provide the necessary template for a major redevelopment project.

A risk management system designed in conformance with ISO 14001 is an essential step in providing a framework for solving the brownfields problem. In a multiple-site project, managers would begin by inventorying and categorizing the properties on the basis of location, size, use (present and past), identity of owners and previous ownership, contamination in soil and groundwater, source of contamination (on-site and off-site), cost of remediation, and future uses. A brownfields database would be the centerpiece for a management program that would be developed and implemented to restore properties or sell them off to businesses or venture capitalists. For those properties that would be retained in the business portfolio, policy planners and senior management can use ISO 14001 to guide their decision making to manage or transfer existing liabilities on the property. A liability and insurance inventory is commonly performed in this process. Customized environmental liability insurance policies, as discussed in the previous chapter, may need to be purchased if properties are to be maintained.

Brownfields joint ventures have been booming in popularity and success in the last three years in the U.S. One example is Cherokee Investment Security, LLC of Denver, CO that is teaming up with several financial partners to make $5 to $500 million deals by either buying businesses or land, or by entering into a risk-sharing joint venture with property owners. Other real estate service firms have joined with environmental remediation firms to jointly propose real estate ventures. In Los Angeles, the California Center for Land Recycling has announced that it will partner with the Trust for Public Land to spur redevelopment of brownfields in that state. The Irvine Foundation has agreed to provide $2 million to finance the startup of that project. For each of these major joint ventures, an ISO 14001 management plan can guide decision making and encourage the development of a cost-effective, risk astute project.

A brownfields redevelopment program with an ISO 14001 component can also fulfill many business objectives. In a recent article, it was suggested that such a program can:

- Prevent redeveloped sites from becoming eyesores again through current or future mismanagement

- Reassure current owners, developers, lenders, and investors that their liability protection will not be compromised by operations of new enterprises
- Offer a new way for brownfields entrepreneurs to achieve environmental compliance without government mismanagement
- Convince government officials and the skeptical public that a trade-off of a lesser level of current clean up for a potentially greater level of environmental consciousness in future site operations is worthwhile (Freeman and Belcamino, 1996).

An important component of ISO 14001 is continual improvement. There is a significant need not to repeat the mistakes of the past. The vast efforts of the federal, state, and local authorities should not be wasted by repeating the mistakes of the 1960s style urban renewal. Environmental issues have become an important part of community awareness and planning. Old approaches to urban development will not work and may become a barrier to successful revitalization of urban projects (Charles, 1997). Three principles are raised in this context by Larry Charles, a noted community activist in Hartford, CT who is in favor of brownfields redevelopment as long as the community (1) is responsible for the success of the project; (2) is allowed to control the project; and (3) is held accountable for the results of the project (Charles, 1997). In short, urban planning cannot be successful without significant involvement by the community which is going to be both responsible for and the ultimate beneficiary of the project.

ISO-Brownfields redevelopment and management projects must not only avoid land-use problems that have created industrial waste, but must also improve the future use of the properties. Energy efficient green buildings, open space, conservation easements, new urbanism, and other innovative planning techniques make important contributions to the quality of life in all communities, but particularly those that have been impacted the most by industrialization. Planning and development must occur with coordination and significant input from transportation experts to make projects pedestrian friendly and promote the use of alternative modes of transportation, such as buses, light rail, van pooling, and bicycles (Kunstler, 1996). Persons familiar with ISO standards are uniquely qualified to provide sound management advice to the communities and property owners, who must work with a range of experts to create projects that will provide sustainability.

Carefully planned efforts will undoubtedly attract environmentally sound industries that will prevent the future spread of brownfields and encourage environmentally and economically sustainable communities. ISO 14000, used in conjunction with economic planning processes, will permit consistency and conformity with accepted standards to replace haphazard industrialization that occurred before people were aware of or cared about the long-term impacts of pollution. Properly utilized, the ISO 14000 standards can assist businesses to remediate brownfields and turn these sites into productive properties that will be continuously improved in the future.

PART TWO:

RISK REDUCTION STRATEGIES

Chapter 10

REDUCING RISKS USING ENVIRONMENTAL POLICY

We begin this section with a look at how businesses have attempted to reduce their environmental risks by adopting their own environmental policy statements. Before the advent of ISO 14000, proactive companies were creating their own policy statements for a number of reasons. An environmental policy, if implemented effectively, can lessen a business's impact on the environment and reduce legal and environmental risks proportional to the degree of commitment by senior management to achieving the goals and accomplishing the objectives of the policy. Direct benefits include improvement in the quality of life for employees and neighbors, upgrading of industry standards, and preservation of the ecosystem. Environmental policy statements have been around much longer than ISO 14000. Therefore, it is useful to examine what caused companies to produce these statements, what are their contents, and how they are implemented to understand how risks can be reduced in the context of making improvements to environmental performance.

The following companies are leaders in the early development of environmental policies. Their statements, and particular approaches to reducing environmental risks, have been reviewed for this chapter:

Company	Year of Policy Examined
3M	1991
American Express	1996
Baltimore Gas & Electric	1993
Baxter International	1996
BFI	1991
Blackstone Valley Elec. Co.	1995
Bristol-Myers Squibb	
British Petroleum	1993
Chevron	
Commonwealth Edison	1996
Concord Resources Group	1993
Conoco	1994
Conrail	1993
Coors Brewing Company	1993
Denver Metro Chamber of Commerce	1995

continued

Company	Year of Policy Examined
Duke Power	
DuPont	1994
Exxon	1996
Florida Power & Light	
General Motors	1992
Geneva Pharmaceuticals	1991
Hauser	
Hewlett-Packard	1993
Hughes Electronics	1995-6
IBM, Intel	1995
Johnson & Johnson	1993
Kimberly-Clark	1990
Lockheed Martin	1996
Monsanto	
Newport Electric	1995
Northern Telecom	
Pacific Gas & Electric	
Philip Morris	
PSI Resources	
Public Service Company of Colorado	1996
Southern California Edison	1993
StorageTek	1995
Sun Company, Inc.	
Sun Microsystems	
Sutton	1995
The Southern Company	
Total Petroleum	1996
Unicom Corp.	1996
Unilever	
Union Carbide	1992
United Technologies	
Volvo Group	
Westinghouse	1992
WMX Technologies	1996
Xerox	1996

BACKGROUND

Companies like Conoco and DuPont originally adopted an environmental policy as early as 1968 and 1971, respectively. IBM and Exxon likewise developed their first environmental policies in 1971. These early statements were prepared before the U.S. created its comprehensive environmental laws and regulations governing their environmental business operations. Environmental risks as we know them today were unheard of during these early days; however, these companies and others were aware of the potential of their operations to degrade the environment and were seeking methods to improve their environmental performance.

In this time frame CBS TV produced a landmark series of reports on pollution in the environment. Momentum was also building in Congress, which passed a host of environmental legislation, including the *National Environmental Policy Act* (NEPA) in 1969, major amendments to the *Clean Air Act* and the *Federal Water Protection Control Act* in 1972, the *Coastal Zone Management Act*, the *Safe Drinking Water Act* and other important statutes. In 1976 Congress passed the *Resource Conservation Recovery Act* (RCRA) to control future handling and disposal of hazardous waste from the cradle to the grave.

Events in the late 1970s, such as the discovery of Love Canal, and other notorious hazardous waste disposal sites, produced the *Comprehensive Environmental Response Compensation and Liability Act* (CERCLA), which was signed into law in December 1980. This act introduced a new legal concept for companies that had disposed of hazardous wastes in the past. Companies could be held strictly liable retrospectively for their past disposal practices and they could be required to clean up all of the wastes at these sites so long as any of their toxic wastes were present. Companies began to realize that they would have to change the manner in which they did business in order to reduce environmental liability risks and impacts.

While the advent of strict environmental legislation was perhaps the primary, but certainly not the exclusive, reason for this change in corporate attitude, other significant forces were at work. Robert Kennedy, the Chairman and Chief Executive Officer of Union Carbide Corporation, described the change that occurred in his company: "In the aftermath of the 1984 Bhopal Tragedy, my predecessor set a goal for Union Carbide to develop a system of safety and environmental management practices second to none" (Kennedy, 1991). William Mulligan, Manager of Environmental Affairs for Chevron, said that over the course of the 1980s he had seen "many polls confirming the importance of the environment to Americans. Only an irresponsible company would dismiss this trend as a passing fad or fail to recognize the need to integrate environmental considerations into every aspect of its business. Environmental excellence has to become part of strategic thinking" (Mulligan, 1991).

According to Robert Kennedy, an environmental communications expert by the name of Peter Sandman identified three stages of the history of corporate environmental communication that directly impacted the creation of comprehensive and effective environmental policy:

- **Stage One** was the entire history of industry until Bhopal, the Rhine River spill, and the Valdez incident. This is called "the Stonewall Stage." We knew what we were doing and we knew what was best for the public. We were making the risk-benefit decisions. We told ourselves that people misunderstood chemical risks or were manipulated by the media and environmental activists.

This stage ended in the mid-1980s when companies realized that stonewalling wasn't working. People who are ignored or misled become angrier, more frightened, and more inclined toward activism. Lawmakers discovered that environmental issues win votes, and they passed new laws — not all of them based on a real understanding of the problems. This led to the next stage.

- **Stage Two** is called "the Missionary Stage." Here we decided to educate the public about chemicals and their risks. We launched programs to teach people that chemicals were safe, contributed significantly to the quality of life, and that chemical plants could be good neighbors. This didn't work.

- **Stage Three**, which is where we are now, is best characterized by initiatives like Responsible Care®. In our industry, we saw that our plants, in some respects, haven't been good neighbors. Most important, we haven't built a good record of open communications with the communities where we operate. The core communication problem in our industry is the need to reduce both industrial hazards and public outrage. To deal with public outrage, we must listen to people. Accordingly, we call this "the Dialogue Stage."

People distrust overnight conversions. But we are now telling people that Responsible Care® is a real change. The public believes we have a poor track record, and we can't change this idea. It doesn't matter who is right or wrong — we have lost that battle. If we admit our imperfections, then we may see a recovery (Kennedy, 1991).

CHEMICAL MANUFACTURERS ASSOCIATION RESPONSIBLE CARE® GUIDING PRINCIPLES:

To recognize and respond to community concerns about chemicals and our operations.

To develop and produce chemicals that can be manufactured, transported, used, and disposed of safely.

To make health, safety, and environmental considerations a priority in our planning for all existing and new products and processes.

To report promptly to officials, employees, customers, and the public, information on chemical-related health or environmental hazards and to recommend protective measures.

To counsel customers on the safe use, transportation, and disposal of chemical products.

To operate our plants and facilities in a manner that protects the environment and the health and safety of our employees and the public.

continued

To extend knowledge by conducting or supporting research on the health, safety, and environmental effects of our products, processes, and waste materials.

To work with others to resolve problems created by past handling and disposal of hazardous substances.

To participate with government and others in creating responsible laws, regulations, and standards to safeguard the community, workplace, and environment.

To promote the principles and practices of Responsible Care®, by sharing experiences and offering assistance to others who produce, handle, use, transport, or dispose of chemicals.

Reprinted with permission from the Chemical Manufacturers Association

In 1987, two years before CERES drafted the *Valdez Principles* and several years before ISO began its environmental management system standards project, Johnson & Johnson released its *Worldwide Statement on the Environment.* In the same year, Polaroid announced its *Toxic Use and Waste Reduction Program.* In February 1989 Baltimore Gas & Electric released its *Environmental Policy Statement.* For these companies and others, environmental policy statements became an integral component of their environmental management systems. They heard the government's, Congress' and the stakeholders' call for a higher level of corporate responsibility for their companies' environmental impacts. Many companies already maintained extensive internal policies, procedures, and guidelines that were used by management and employees in conducting systems operations, but had never drafted a specific statement of environmental principles for public release. Others, like 3M, had been working with environmental policies for years. In 1991, 3M released an environmental *Progress Report* which contained the "Corporate Environmental Policy" it developed for its 89,000 employees and facilities in 52 countries. 3M stated that since 1975 it has adhered to an internal policy to discover innovative ways to "lessen [its] impact on the environment — in areas as diverse as product manufacturing processes and recycling programs."

In April 1992 General Motors issued its *Environmental Principles* stating that its compilation of environmental principles represented "an affirmation and summary of policies that, in some cases, date back to the 1950s." General Motors, and other businesses, decided to draft environmental policies that combined, in one document, workplace principles and corporate ethics that had guided the company's operations for many years. 3M's and GM's policies are by no means just a restatement or summary of previous internal policies and procedures condensed into an environmental policy. They are oriented towards establishing future preeminence in the environmental field and reducing risks.

Prior to ISO 14001, most environmental policies have been created for manufacturing and heavily industrialized companies in the petroleum, paper, automotive, transportation, defense, waste disposal, computer, chemical, pharmaceutical, brewery, and gas and electric utility industries. These companies were frequently defendants in massive environmental litigation and, accordingly, were highly motivated to improve their systems' operations to reduce the likelihood of further environmental liability. A review of their policies reveals one of the basic reasons why so many companies decided to draft their policies to address their individual needs.

Each of these industries is vastly different, as are the representative companies within them. The companies have widely diverse operations, products, services, and individual impacts on the environment. They needed to take a comprehensive look at their immediate and long-term impacts on the environment and make a calculated business decision as to how much they needed to improve their environmental performance.

ISSUES IN DRAFTING A POLICY

Drafting an environmental policy statement for these companies turned out to be a rigorous experience that required a thorough knowledge of the entire business. Environmental policies were geared to the nature, scale, and environmental impacts of the company's activities, products, or services. There was considerable initial study to determine the realistic environmental goals and objectives of each of the businesses. Responsible managers evaluated their company's past environmental performance with regard to RCRA, CWA, CERCLA, and CAA, as well as environmental compliance issues, citizens' suits, and other potential environmental liabilities. Certainly, they also measured the current and foreseeable environmental impacts of the business. They addressed the specific environmental health and safety concerns of the company, taking into consideration the needs of the employees, the consumers, and the general public. They determined how much effort was needed to proceed with the implementation of the policy and delegated that responsibility to appropriate employees.

Companies need to have a great deal of latitude in designing a policy that is structured to achieve individualized corporate objectives. Some may want to demonstrate uncompromising commitment to reducing pollution by utilizing the latest technology to reduce emissions and wastes. If ISO 14000 is the approach that is taken, a company must commit to continual improvement. Others simply may need the opportunity to plan an effective policy and to develop an implementation program that can be periodically revised and upgraded to account for business developments, including improvements and set backs, employee training and experience, and unforeseen problems.

Given our rapidly expanding universe of knowledge regarding the environment, it makes sense that companies be permitted to amend outdated policies in an effort to improve environmental performance. Conoco, a pioneer among the companies that have adopted policies, has revised its environmental policy six times since 1968. When contacted for our review, several other companies indicated that their policies are currently being revised and updated. Changing regulations, marketplace conditions, and business practices are all factors which require businesses to amend their policies as needed. For these businesses, there is no choice between accepting someone else's idea of what their principles should look like, or designing their own principles to meet their present and foreseeable future needs.

Corporate Leadership

Companies that have uncompromised commitment from senior management are the most likely to have effective implementation of their environmental policies. A recent survey on what causes corporate environmental responsiveness concluded that attention and support of corporate leadership was critical to the success of the environmental management system (Judge, Miller, and Fowler, 1996). Other factors that cause corporate environmental responsiveness include:

- The degree which managers perceive social expectations concerning the natural environment to be legitimate
- The availability of sufficient financial resources
- The degree of cooperation with regulatory agencies to produce *win-win* relationships
- The degree to which managers perceive environmental regulations to be ambiguous, uncertain, and inconsistent
- The perception of the cost of compliance

If the board of directors and senior management perceive all of these factors to warrant unqualified support for improvements in environmental performance, then the company is well on its way to an effective environmental management system that will ultimately reduce risks. The degree of commitment from senior management is often demonstrated in how the policy is released to the employees and general public and then how it is implemented.

Company boards of directors often adopt their companies' policies (WMX Technologies, Inc., Kimberly-Clark, Concord Resources Group, and Coors Brewing Company). This demonstrates that an environmental policy is an important corporate objective. Some of the policies are officially released to the general public by the president, CEO, or other responsible management official in the company who issues a declaration or statement regarding the policies (Conoco, Johnson & Johnson, Geneva Pharmaceuticals, Intel, and Total Petroleum). This demonstrates accountability of senior management for environmental policy commitments.

Objectives and Goals

Many of the policies are preceded by a special preamble or introduction that sets forth the key objectives of the business in protecting the environment and the health and safety of its employees, consumers, and the general public. For example, Hewlett-Packard states that it "is committed to conducting its business in an ethical and socially responsible manner. An aggressive approach to environmental management, which includes occupational health, industrial hygiene, safety management and ecological protection, is consistent with the spirit and intent of our established corporate objectives and cultural values." Commonwealth Edison, in its preamble, states "We have a strong incentive to be sensitive to environmental issues...We believe that our customers want and expect us to respect the environment...As members of the communities in which we operate, Edison's employees are also deeply interested in the effects our operations have on the environment, as well as on the public health and safety in our service area." Xerox's CEO and Director state in their company's policy preamble that "mounting concern for the environment and the recognition that sustainable development is essential to our common future are key components . . .". These preambles set the tone and demonstrate the legitimate commitment of the leadership of each company to conduct its operations as a responsible steward of the environment.

Some businesses set specific objectives or goals that they want to achieve when their policy is fully implemented. Westinghouse, for example, has declared that it will "[r]educe toxic chemical releases to the air 50% by 1995 and 90% by 2000 based on 1988 levels." By publicly announcing this goal as an integral component of its environmental policy, the leadership of Westinghouse has made a substantial public commitment to moving beyond compliance. Likewise, Baxter has agreed to establish and maintain a state-of-the-art environmental program throughout the world. Other businesses have joined Westinghouse and Baxter in announcing similar goals of waste reduction.

Both Westinghouse and Baxter engaged in extensive strategic planning and careful study of their capabilities to reduce environmental impacts before setting these goals. Both companies dedicated a substantial amount of time and resources to achieve these milestones. Other businesses that cannot afford to make such capital investments should avoid setting unreasonable standards or unachievable goals because failure may increase the possibility of shareholder dissatisfaction, public censure, and consumer confusion. For these reasons, policy statements must be worded carefully to meet realistic expectations and to assure that reasonable corporate objectives are accomplished. Otherwise, claims of *greenwashing* can occur which can damage corporate reputations and negatively impact market share.

Ill-conceived policies can also undermine management and result in employees, shareholders, and the general public losing confidence in corporate commitment to the environment. The purpose of an environmental policy is to prevent environmental controversy, not foment it. In a recent survey conducted by the Council of Environmental Priorities (CEP), one large business was given a "poor environmental performance" in the group's annual rating of corporations because of a higher than industry average of OSHA violations. The company's environmental policy from the previous year had indicated a commitment to "avoid[ing] unacceptable risk to human health and safety." Businesses which fail to meet the aggressive goals and principles in their policies may diminish any positive gain from having the policy in the first place.

Many of the businesses incorporated in their policy statements several of the key environmental concepts contained in the *CERES Principles*. For example, most of the businesses would readily agree that their policies endorse protecting the biosphere; sustaining renewable resources and conserving nonrenewable resources; reducing waste; wisely using energy; reducing risks; marketing safe products and services; and disclosing safety hazards to employees, consumers, and the general public. It is apparent that these companies preferred to set individualized standards that represent each business's commitment to respect and protect the environment.

How Many Principles are Enough?

The length of environmental policy statements in our review differs dramatically. For example, some companies' environmental policy statements set forth a list of company principles involving environment, health, and safety issues. These companies decided to keep their policies concise without including any descriptive detail concerning past, present, or future performance. Others list their company principles and illustrate them with examples of corporate practices and programs that have successfully reduced waste and had other beneficial environmental impacts. Some actually give examples of quantitative data, detailing some of their environmental successes. The wide range and number of principles adopted by the companies and the differences in the style of writing policy statements demonstrate that there is no uniform approach to drafting a policy. A company that adopts a greater number of principles is not better or more environmentally correct than another company that has endorsed fewer principles. Each business implements its company's principles in a different manner based upon individual needs. Many of the companies' principles contain the same or overlapping concepts, and the principles may include one or more environmental concepts. Accordingly, the number of principles adopted bears no relationship to the quality of the environmental compliance policy.

Corporate Health, Environment and Safety Concepts

The concepts contained in the companies' environmental policies we have reviewed have been organized and divided into five categories: general concepts, consumer protection, internal controls, employee protection, and community protection. The following is a brief description of these concepts together with representative examples from particular companies' policy statements:

General Concepts

<u>Safe waste disposal</u> means to provide for the legal and safe disposal of wastes generated by the business and to comply with all environmental regulations for the storage, transportation, and disposal of all waste. Example, Blackstone Valley: "Will make every effort to minimize the creation of waste, especially hazardous waste, and dispose of wastes through safe and responsible methods."

Efficient use of energy means to use electricity and fossil fuels in the most efficient manner possible and to promote use of newer technology and methods that maximize output while minimizing energy consumption. Example, Chevron: "Conserve Company and natural resources by careful management of emissions and discharges and by eliminating unnecessary waste generation. This also includes wise use of energy in operations."

Reduction of common forms of pollution means to minimize releases of water, air, toxic, and atmospheric pollution. Some examples are pledges to reduce air or chlorofluorocarbon emissions by a certain percentage, or better water and air purification systems. Example, Coors Brewing Company: "We apply innovative technology toward the efficient use of resources, including reduce/reuse/recycle actions, waste minimization and the reduction of emissions to air, water and land."

Self-regulation and auditing means to monitor and to evaluate environmental performance by gathering information on safety, health, and environmental quality. Examples, IBM: "Conduct rigorous audits and self-assessments of IBM's compliance with this policy, measure progress of IBM's environmental affairs performance, and report periodically to the Board of Directors." Johnson & Johnson: "To insure universally high standards, we instituted a periodic environmental assessment program by independent outside reviewers."

Environmental responsibility means to take responsibility for the company's actions and impacts on the environment. This is achieved by acting responsibly regarding the disclosure of information, the discard of waste, and other aspects of business that may affect the environment. Some policies provide for restoration of the environment for any negative impacts or accidents that harm the environment. Examples, Hewlett-Packard: "Proactively address environmental contamination resulting from any HP operation." Bristol-Myers Squibb: "…work in an integrated manner to identify, evaluate, and resolve EHS [Environmental, Health and Safety] impacts related to the management of resources and their related byproducts — i.e., selection, use, and exposure."

Sustainability means to recognize environmental management as among the highest corporate priorities; and to establish policies, programs, and practices for conducting operations in an environmentally sound manner. Examples, Xerox: "Protection of the environment… from unacceptable risks takes priority over common considerations and will not be compromised." Public Service Co. of Colorado: "… continually search for ways to improve the performance of our environmental protection and safety systems." Baxter: "We will strive to conserve natural resources and minimize or eliminate adverse environmental effects and risks associated with our products, services and operations." WMX Technologies, Inc.: "The Company will use renewable natural resources, such as water, soils and forests, in a sustainable manner and will offer services to make degraded resources once again usable." Unicom Corporation: "Encourage waste reduction, by-product marketing, recycling, re-use, sale, salvage, and other means of supporting the life cycle management strategy."

Compliance with all laws and regulations means to comply with all relevant and applicable environmental laws and regulations. Some businesses also strive to go "beyond compliance" with existing laws and regulations, and to develop more technologically sound and economically responsible laws and regulations. Examples, Baltimore Gas and Electric: "We are committed to conducting our business in compliance with the letter and spirit of the law. Where appropriate, we establish our own policies and standards which go beyond what is legally required. Our goal is to be a model of effective environmental responsibility and a well-regarded corporate neighbor." Exxon: ". . . apply responsible standards where laws and regulations do not exist ... work with government agencies and others to develop responsible laws. . .".

Conservation of nonrenewable resources means to attempt to minimize the use of nonrenewable resources such as fossil fuels. This can be accomplished through greater efficiency and updated equipment, or through improvements in production. Example, Xerox: "All Xerox operations must be conducted in a manner that safeguards health, protects the environment, conserves valuable materials and resources and minimizes risks of asset losses."

Minimize waste means to keep the production of waste, especially needless waste, to a minimum by attempting to maximize recycling and conservation. Examples, Monsanto: "Search worldwide for technology to reduce and eliminate waste from our operations, with the top priority being not making it in the first place." DuPont: "Goal of zero waste and emissions." Storage Tek: "... strive for continuous improvement in pollution prevention/waste minimization and resource conservation and will monitor performance against defined goals."

Recycle and promote use of renewable resources means to implement recycling and renewable resource programs. Examples, Hewlett-Packard: "Design our products and services and their associated manufacturing and distribution processes to be safe in their operation; minimize use of hazardous materials; make efficient use of energy and other resources; and to enable recycling and reuse." PSI Resources, Inc.: "Pursue methods to prevent pollution and conserve raw materials, including recycling waste and promoting the efficient use of energy by our customers through all cost-effective means."

Consumer Protection

Promote and market safe products means to provide consumers with products which pass all applicable safety codes and regulations and advertise products and promote the use of products in the safe way in which they are intended to be used. Examples, DuPont: "Will determine that each product can be made, used, handled and disposed of safely and consistent with appropriate safety, health and environmental quality criteria." Volvo Group: "Seeking to ensure that production processes and products comply with comparable environmental standards, wherever in the world the company operates."

Consumer education means to provide consumers with enough information and training to insure safe use of the products. This may be accomplished through seminars, workshops, publications, instruction manuals, and advertising. Example, Chevron: "Counsel customers, transporters and others in the safe use, transportation and disposal of raw materials, products and waste."

Product stewardship means to uphold a published product code which can entail a variety of specifications which may include product testing techniques, employee standards and input requirements. Example, Westinghouse: "Protect and enhance our customer's environmental reputation by developing and providing environmentally-sound products and services."

Compliance with all consumer product laws and regulations means to adhere to all laws that protect consumers. This includes special safety requirements and product recall responsibility. Example, 3M: "Assure that its facilities and products meet and sustain the regulations of all federal, state and local environmental agencies."

Safe packaging and transportation of products means to package products in a way that provides protection to both consumers and the general population and to transport products safely to protect the public from inherent risks. Examples, British Petroleum: "All products that BP makes for sale or use, and products rebranded by BP, will be evaluated to ensure that, despite inherent hazards, they can be stored, handled, transported and used safely." Volvo Group: "Working to develop efficient transport systems having minimum environmental impact."

Safe and adequate testing of products means to test new products sufficiently while minimizing the negative impact on those tested (humans and animals). Example, Sun Company: "Through research, planning and analysis we will strive to minimize the environmental impact and health and safety hazards of our business products and services."

Internal Controls Concepts

Environmental leadership means to lead the company, and the industry, in a responsible manner to achieve health, environmental, and safety goals. Example, Baxter: "Will work to become a leader in respecting the environment. Environmental excellence is vital to Baxter's business interests and is consistent with our mission and shared values."

Cooperation within the industry for progress means to agree to work closely with other competitors in the industry for the benefit of the environment and to share information to reduce the company's negative environmental impacts. Examples, IBM: "Assist in the development of technological solutions to global environmental problems, share appropriate pollution prevention technology and methods, and participate in efforts to improve environmental protection and understanding throughout the industry." Pacific Gas and Electric Company: "Work cooperatively with others to further common environmental objectives." Sun Microsystems: "Incorporate environmental criteria in vendor, supplier and strategic partner relations ... share 'Best Available Technology.'"

Environmental research and development means to research and develop new techniques and systems that benefit the environment. Examples, Duke Power: "We will support research aimed at enhancing our knowledge of the environment and minimizing environmental impacts of power generation. We will share this information with others." Unilever: "Take an active part in protecting the environment through continuous improvements in the environmental impact of their operations."

Toxic and hazardous waste policies means to have a specific policy for contact with toxic and hazardous wastes, a safe transportation and storage plan, and warnings of specific health risks. Examples, Newport Electric: "Will make every effort to minimize the creation of waste, especially hazardous waste, and dispose of wastes through safe and responsible methods." Westinghouse: "Reduce toxic chemical releases to the air 50% by 1995, and 90% by 2000 based on 1988 levels."

Investment for improved energy efficiency means to spend money for newer updated equipment which uses less energy, such as buying newer fuel-efficient vehicles, lower-wattage light bulbs, and equipment that requires less energy input to complete the same task. Examples, Blackstone Valley: "We will invest in energy-efficient equipment and practices at our own facilities, and we will promote the efficient use of energy by our customers through education, promotion and investment in conservation and load management measures." Johnson & Johnson: "... assuring wise use of resources and minimizing waste ... [by] participating in a ... pollution prevention program that is designed to reduce pollution by cutting national electricity demand by ten percent or more."

Risk minimization and accident prevention means to set up a program to educate and warn employees, consumers, and the community of all possible health, environmental, and safety risks. Examples, Coors Brewing Company: "We strive to minimize environmental, health and safety risks to our employees and the communities in which we operate. We prepare for emergencies and communicate appropriately and responsibly." Blackstone Valley: "BVE minimizes the health and safety risks for our employees, customers, and the communities in which we operate by employing safe work practices and procedures and by being prepared for emergencies."

Crisis management means to adopt a plan which sets forth the steps to take if an accident occurs. The plan should include specific guidelines and a chain of command to immediately respond to a crisis. Examples, DuPont: "Will inform employees and the public about the safety and health effects of its products and workplace chemicals and will provide leadership in establishing programs to respond to emergencies involving hazardous materials in communities where the Company has a significant presence." Conoco: "Maintaining emergency preparedness plan and response capabilities."

Damage restoration means to restore the environment to a suitable state for companies which, due to the nature of their business, alter the environment. Examples, WMX Technologies: "The Company will take responsibility for any harm we cause to the environment and will make every reasonable effort to remedy damage caused to people or ecosystems." Commonwealth Edison: "When we are responsible for spills and accidental releases of contaminants, it is our policy to acknowledge that responsibility ... we will ... promptly make all necessary notifications to appropriate authorities ... promptly and thoroughly remove contaminants to designated levels ... investigate and clean up contamination to the environment caused by those past operations for which we are responsible."

Corporate leadership and responsibility means to have a member of top management, such as a vice-president or member of the Board of Directors, involved in health, environmental, and safety issues. This individual should be well educated and informed regarding environmental impacts, and should also be responsible for decisions made which have an effect on the environment. Examples, Lockheed Martin: "The Company President is responsible for compliance with corporate, federal, state and local EHS requirements, and shall pursue opportunities to prevent workplace injuries and illnesses at all operating locations and to enhance environmental quality." American Express: "The Public Responsibility Committee of the Company's Board of Directors will oversee management's commitment to environmental policies and practices."

Operating line organization and responsibility means assigning accountability and responsibility for attaining compliance with environmental requirements. Examples, Lockheed Martin: "Establishing corporate-wide EHS policies and functional procedures that define the Corporation's requirements and assign accountability to appropriate levels." BFI Waste Systems: "Require each business segment to develop specific plans, programs and procedures appropriate to that segment to ensure effective implementation of this policy." Commonwealth Edison: "Each line organization is responsible for complying with our environmental policies in its own operations."

Political activity and input means to be an informative and well-rounded source of information on environmental issues which affect the industry and to maintain an active role in the evolution of environmental policy and regulation through assistance of federal, state, and local agencies. Examples, 3M: "Assist, wherever possible, governmental agencies and other official organizations engaged in environmental activities." General Motors: "We will continue to work with all governmental entities for the development of technically-sound and financially-responsible environmental laws and regulations."

Compliance of agents means to ensure that those persons acting on behalf of the company, such as suppliers and contractors, comply with the company's environmental policy. Suppliers and contractors includes those in other countries, for companies that conduct international business. Examples, Bristol-Myers Squibb: "… give preference to suppliers and contractors whose EHS commitment and practices are consistent with its own, and who have demonstrated environmentally responsible products, services, and management." StorageTek: "… encourage its suppliers and contractors to adopt the principles in this policy."

Employee Protection Concepts

Environmental training and education means to adequately train and educate all employees in both safety in the workplace and for environmental safety; to explain hazards in the workplace and potential hazards to the environment due to possible accidents; and to provide information that directs the employee in basic crises management, accident prevention, and risk minimization. Examples, Blackstone Valley: "Minimize the health and safety risks for our employees, customers, and the communities in which we operate by employing safe work practices and procedures and by being prepared for emergencies. Employees will receive adequate training, tools, and management support to perform their environmental responsibilities." Pacific Gas and Electric Company: "Provide appropriate environmental training and educate employees to be environmentally responsible on the job and at home."

Employee responsibility means to make the employee responsible for following all health, environmental, and safety standards set by the business and by regulation. This places a duty on each employee to be personally responsible for safety in the workplace. Examples, Coors Brewing Company: "We have a systematic program of employee education and all employees are responsible for adhering to these principles." Exxon: "It will encourage concern and respect for the environment, emphasize every employee's responsibility in environmental performance, and ensure appropriate operating practices and training."

Employee disclosure means to disclose all risks or hazards to employees and have in place policies which allow employees to come forward with health, environmental, and safety concerns. Examples, Union Carbide: "Report promptly to officials, employees, customers and the public, information on chemical-related health or environmental hazards and to recommend protective measures." Newport Electric: "[The Company] will disclose to its employees and to the public any incident relating to our operations that causes environmental, health or safety hazards."

OSHA and ADA compliance means to follow and comply with all the standards set forth by the Occupational Safety and Health Administration (OSHA) and the *Americans with Disabilities Act* (ADA). Example, American Express: "Comply with federal, state and local laws, e.g., Occupational, Safety and Health Administration (OSHA) guidelines."

Self-regulation policy for safety means to have an internal body for researching, inspecting, and evaluating health and safety in the workplace. Example, Exxon: "Conduct and support research to extend knowledge about the safety effects of its operations, promptly applying significant findings and, as appropriate, sharing them with employees, contractors, government agencies and others ... undertake appropriate reviews and evaluations of its operations to measure progress and to ensure compliance with this safety policy."

Community Protection Concepts

Company tours means allowing the public directly into the workplace. Company tours provide a direct link with the public and exhibit a feeling of openness toward the community. Examples, Concord Resources: "In order to allow for exchange of information with communities near our facilities, we will conduct public tours, consistent with safety practices." Monsanto: "Keep our plants open to our communities and involve the community in plant operations."

Communication and information means to communicate responsibly to the community about the operations of the company and how it may affect the surrounding community. Examples, British Petroleum: "We will communicate openly with those who live or work in the vicinity of our facilities to ensure their understanding of our operations, and our understanding of their concerns." Concord Resources: "Concord will encourage open dialogue with the public. We wish to assist the public in understanding the environmental impacts of our activities by providing information." Baltimore Gas and Electric: "Respond to our customers, neighbors, employees, regulators and others whenever they have concerns about the environmental impacts of our business." Sun Microsystems: "Share program successes with employees, customers, and the general public to further the efforts of environmental stewardship."

Special programs means to adopt special programs for enhancing the environment. Some examples are adopt-a-highway, wetlands protection/formation, or adopt-an-animal programs, which are for the sole benefit of the community and the environment. Example, Florida Power: "Maintain active programs for protecting endangered species such as the American Crocodile, the Florida manatee, and the southern bald eagle. The company has preserved under its stewardship, the Barley Barber Swamp ... and promotes conservation of endangered sea turtles ..."

Cooperation with special interest groups means to work together with interest groups in achieving the common goal of protecting and preserving the environment. Examples, Concord Resources: "Concord will also support and participate in the development of public policy and in educational initiatives through cooperation with government, environmental groups, schools, universities and other public organizations which will assist in protecting human health and the environment." WMX Technologies, Inc.: "The Company will encourage its employees to participate in and to support the work of environmental organizations, and we will provide support to environmental organizations for the advancement of environmental protection." PSI Resources: "Develop and maintain open and constructive relationships with environmental groups, regulatory agencies, public officials, business and residential customers, employees and concerned citizens." Hughes Electronics: "... create and support forums to facilitate information transfer." Dupont: "... build alliances with governments, policy makers, companies and advocacy groups to develop sound policies ...".

How are the Policies Implemented?

Many businesses go beyond espousing principles in their policy statements by addressing in the policies themselves how they actually will be implemented. By setting forth the means by which the goals and objectives will actually be accomplished, businesses assume additional responsibility for their environmental actions. Such open displays raise the expectations of both stakeholders and the general public that the company will achieve these goals. Several of the companies that were reviewed appointed senior corporate officers or committees to oversee the implementation of the policies or have developed internal compliance procedures and programs to direct compliance efforts and to evaluate environmental performance. The following is a summary of how 12 businesses describe in their statements how they will implement their policies:

Johnson & Johnson
When Johnson & Johnson issued its *Worldwide Statement on the Environment,* it declared that "[t]he managements of all companies, operating units, manufacturing plants and research laboratories are expected to ensure their operations and facilities meet the requirements of the Corporate policy and guidelines, as well as all applicable laws and government regulations." The corporate staff was ordered to assist operating companies in the implementation of this policy by developing appropriate guidelines and procedures; establish and maintain liaison with appropriate governmental agencies; provide training and consultation

to operating company personnel; audit facilities for compliance; and report annually on environmental issues, programs, and compliance.

Baxter

Baxter implemented its *Environmental Policy* by having its Public Policy Committee of the Board of Directors appoint an Environmental Review Board (ERB) to oversee implementation. "The ERB will review and decide matters of environmental importance and will make an annual report to the Board of Directors." Further, Baxter required "[t]he manager of each manufacturing and distribution facility, and other division group managers where appropriate, will appoint a qualified environmental representative to coordinate and manage the unit's environmental program." That manager is responsible: "to determine the facts regarding generation and release of pollutants from its facilities and responsibly manage its affairs to minimize any adverse environmental impact; to develop and implement an environmental management program to comply with the Policy; to select, design, build, and operate products, processes, and facilities in order to minimize the generation and discharge of waste and other adverse impacts on the environment; and to utilize control and recycling technology wherever scientifically and economically feasible to minimize the adverse impact on the environment."

WMX Technologies, Inc.

As part of its implementation strategy, WMX agreed to prepare and make public an annual report on its environmental activities. This report includes a self-evaluation of the company's implementation of its principles, including an assessment of the company's performance complying with all applicable environmental laws and regulations throughout its worldwide operations. Like Baxter, WMX created an Executive Environmental Committee, which includes environmental professionals. That committee reports directly to the Chief Executive Officer regarding the implementation of this policy.

Total Petroleum, Inc.

On August 30, 1993 Total, like Baxter and WMX, announced that its environment, safety, and public affairs department would inform and advise the executive committee and the Board of Directors on environmental, safety, and health issues, as well as assist and advise the operating units of the corporation in the implementation of the principles. This department would also serve as the corporation's "focal point" for all external agencies on environmental, safety, and health matters. Total also noted that its executive committee would be responsible for issuing specific policies that expand the provisions of its principles and provide overall direction for corporate-wide implementation of environmental, safety, and health principles.

BFI

BFI's *Environmental, Health and Safety Policy* contained eight steps to implement the policy: (1) conduct appropriate training and audit programs; (2) proactively identify and control hazards to health, safety, and the environment resulting from its operations; (3) conduct appropriate information sharing programs to communicate the significant operating aspects of the company's facilities with employees, the surrounding communities, and appropriate regulatory agencies; (4) utilize cross-company quality committees to identify and develop, where appropriate, additional business environmental, health, and safety policies which are more protective than existing laws and regulations; (5) encourage those affiliations where the company would not be the majority owner to adopt policies comparable to its environmental, health, and safety policy; (6) require each company segment to develop specific plans, programs, and procedures appropriate to that segment to ensure effective implementation of this policy; (7) work constructively with trade associations, elected officials, governmental agencies and others to develop equitable and effective laws and regulations to protect human health and the environment; and (8) conduct reviews of new facility designs and construction specifications to assure that appropriate environmental, health, and safety controls are in place.

Coors Brewing Company

On August 12, 1993 Coors announced that to implement its environmental occupational health and safety policy, adopted on the same date by the Board of Directors, it would develop "an effective program of measuring progress in implementing these principles. Our progress is measured by attainment of specific objectives and goals which seek continual improvement."

Lockheed Martin

Lockheed Martin required that the "presidents and heads of operating elements are responsible for ensuring all corporate-owned or managed facilities comply fully with federal, state and local environmental laws and regulations." The "corporate environmental management will establish environmental policy and issue environmental criteria for the corporation." Additionally, "operating elements will institute pollution prevention programs aimed at substantial reductions in the use of hazardous substances and in the generation of waste and pollutants, and the corporate environmental management vice-president will issue directives and procedures to implement the policy."

Commonwealth Edison

Commonwealth Edison issued its *Environmental Policy Statement* that requires "each line organization [to be] responsible for complying with our environmental policies in its own operations." The environmental service department is responsible for assuring that the company adheres to the policies. An executive committee, under the chairmanship of the president, establishes environmental policy and monitors environmental compliance for the company. A committee of the Board of Directors has been selected to oversee environmental policies of the company."

Chevron

Chevron has agreed to implement its *Environmental Principles* by conducting "a comprehensive compliance program including audits."

IBM

IBM's *Corporate Policy 139, Environmental Affairs* provides that IBM will "conduct rigorous audits and self-assessments of IBM's compliance with this policy, measure progress of IBM's environmental affairs performance, and report periodically to the Board of Directors."

Duke Power

Duke Power assigned environmental activity oversight responsibility to a senior business executive. Like Chevron and IBM, Duke Power agreed to "perform periodic reviews and audits to assure effective programs and practices that are consistent with its principles," which are entitled *Duke Power's Commitment to the Environment.*

Halliburton Company

The company's environmental commitment is implemented through its Environmental Quality Improvement Process (EQIP). This program is administered by the chief environmental officer, a member of top management designated by the company's CEO, with the oversight of the Environment, Health and Safety Committee of Directors. The chief environmental officer oversees the implementation of the environmental policies and procedures; administers EQIP; recommends to the Environment, Health and Safety Committee annual goals for such a program; and reports on environmental matters and compliance with the policy to the EHS Committee at each of its meetings and to the executive committee whenever appropriate. The Environmental Quality Task Force assists the chief environmental officer in the administration of EQIP and, under such officer's direction, develops, coordinates, and monitors specific programs to promote and implement the EHS policies and procedures. Each operating group also must adopt a policy substantially equivalent to this corporate policy.

CONCLUSION

These 12 examples demonstrate that implementing an environmental, health, and safety policy is far more difficult than simply drafting the policy. Businesses must recognize that without careful planning at the outset, implementation can be a frustrating and expensive experience — particularly where environmental principles conflict with practical and potentially less expensive solutions. For these reasons, businesses should rely on top management or a committee under the supervision of the board of directors or the executive officers to guide the process, and delegate to mid-level management the responsibility to draft additional instructions and guidance to achieve corporate environmental, health, and safety objectives.

In Chapter 5 we discussed the benefits and costs in developing an ISO 14001 environmental management system, of which an environmental policy is a key component. Here it is important to review briefly the benefits and costs of developing an environmental policy. A policy obviously should never be viewed except in the context of the environmental management system that is built around it. The policy, however, is a public document and companies need to understand that it will be read frequently by itself, without the implementation plan or the remainder of the environmental management system read in context.

Businesses that take the time to develop an effective environmental policy are making a wise choice. Certainly, there are significant incentives and rewards for those companies that decide to move beyond environmental compliance. Undoubtedly, there are also costs. When businesses use only safe and recyclable materials in manufacturing, conduct exhaustive independent environmental audits, and use suppliers who meet or exceed their own environmental standards, operating costs can be expected to increase. A business must also live up to its self-defined policy concepts. This is often quite costly and to fail in this endeavor risks its reputation. Although there are no specific environmental cases on this particular point, a business that voluntarily assumes a duty beyond that required by law may be required to live up to that duty.

When a company designs an environmental policy and implements an environmental management system effectively, the results may be difficult to measure in the short term. It is not easy to calculate, for example, the exact amount of money that businesses have saved. The real savings may be lawsuits that were never filed, government investigations that were never conducted, and catastrophic events that did not occur. Intangible environmental benefits unquestionably outweigh the costs of preventive environmental practices.

While it is difficult to measure costs, it is easy to measure other types of benefits. Progressive environmental policies have received unexpected economic benefits in the forms of reduced expenses for waste removal and added revenues from the sale of recycled materials. Businesses that adopt an environmental policy are making a strong public declaration to their employees, their customers, the general public, and their competitors that they care about their impact on the environment and are willing to devote corporate time and resources to improve performance and to solve problems. Consumers are starting to shop corporations instead of brands. Environmental policy statements written with the guidance of ISO 14000 can convey an effective message that the company's and public's concerns are co-extensive, and that each has a duty to strive to safeguard, protect, and sustain natural resources.

Further, a thorough, balanced policy can be convincing proof that a business is a good corporate citizen and not delinquent about environmental compliance efforts. In the event of a government investigation, an environmental policy that has been implemented effectively and without compromise can be used to convince the government that a regulatory violation was an isolated aberration, and contrary to stated corporate policy. The violation, therefore, should be treated as a civil rather than a criminal offense.

A company's biggest mistake is to adopt a paper policy with no intention of implementing the principles. The draft *U.S. Sentencing Guidelines for Environmental Offenses* provide that fines and penalties for environmental compliance will be substantially increased if the government discovers that a defendant-company adopted a compliance policy in name only. Even without the specter of a government criminal investigation, businesses can ill afford to misrepresent their environmental record or their commitment in an effort to persuade their employees, stakeholders, customers, or the general public that they have a program which in reality does not exist.

Companies that effectively implement their environmental policies will be more likely to identify the causes of noncompliance and, hopefully, will reduce the likelihood of violation of environmental laws and regulations. Reducing risks is the key advantage of creating an environmental policy. This comes with a significant cost. Companies with effective environmental management systems, including auditing programs, may discover that they have violated environmental laws. These companies will be more likely to disclose the violation to the government if they have a strong environmental policy that states an uncompromised commitment to abiding by the terms of relevant environmental statutes and regulations. The decision to disclose is not automatic, based upon the existence of a policy. We know of no company that has a policy to disclose any and all violations to the government. The decision to disclose is complicated, as are the consequences that follow, both of which are the subjects of the next chapter.

Chapter 11

REDUCING RISKS BY AUDITING AND DISCLOSING VIOLATIONS

The EPA wants companies to conduct environmental audits and to disclose violations in order to avoid criminal liability and stiff, gravity-based penalties. Gravity-based penalties are defined as "that portion of a penalty over and above the economic benefit; i.e., the punitive portion of the penalty, rather than that portion representing a defendant's economic gain from noncompliance." This chapter explores how the EPA's new policy works and its relationship to the states that have co-extensive jurisdiction to prosecute environmental violations. The EPA's policy and the audit privilege statutes enacted by many states present new opportunities for businesses to create sound environmental management systems that reduce the risk of noncompliance.

On January 22, 1996 the EPA commenced a new audit policy which is intended to offer businesses practical and legal incentives to conduct comprehensive environmental audits and create environmental management systems. When violations are discovered through these programs, businesses can reduce their risks and liabilities by self-reporting. The EPA's policy is controversial because many businesses believe that if they conduct audits, and find and correct regulatory noncompliance, they should not be penalized for their proactive efforts to solve environmental problems. Many states agree with businesses on this issue and have passed legislation encouraging businesses to report noncompliance and obtain total or partial immunity from penalties. The existing federal-state conflict underscores the evolving nature of environmental regulations and demonstrates the need for businesses to control their own destinies by implementing effective risk management control systems to prevent noncompliance from occurring.

In the first year since the policy took effect, 105 regulated entities have disclosed environmental violations at approximately 350 facilities. The EPA has settled with 40 businesses, involving violations at 48 facilities and waived penalties in most of these cases. Companies that have self-reported environmental violations include Alyeska Pipeline, Baldwin Piano & Organ, General Electric Corp., Minolta Co., and Sunbeam-Oster Co.

General Electric discovered, disclosed, and corrected violations of the *Clean Air Act* at its silicone manufacturing facility in Waterford, NY. According to the EPA, the U.S. Department of Justice and the EPA waived a gravity-based penalty, which reduced the actual penalty to $60,684 and reflected the amount of economic benefit the company gained from noncompliance. Vastar Resources,

Inc. discovered, disclosed, and corrected *Clean Air Act* violations and reduced its fine to $137,949 after the gravity-based fine of several hundred thousand dollars was waived. Cenex, Inc. reported its failure to file reports under the *Toxic Substances Control Act,* received a penalty reduction of $318,750, and paid a fine of $106,250.

These cases demonstrate that EPA is willing to reduce, but not waive, fines altogether when the audit policy is invoked. Steve A. Herman, Assistant Administrator of EPA's Office of Enforcement and Compliance Assurance, has said that environmental auditing can be encouraged without blanket amnesties or audit privileges that would excuse serious misconduct, frustrate enforcement, encourage secrecy, boost litigation, or lead to public distrust. Herman's office issued an interpretative guidance on January 15, 1997 explaining how the audit policy is applied to the regulated community.

If a business uncovers a violation using either environmental auditing or a compliance program, the policy provides incentives to correct the problem and disclose the violation voluntarily to the government to avoid civil and criminal penalties. The policy makes no guarantees, however, that the business will escape all liability. General Electric, Vastar, Cenex, and a host of other companies have discovered that their proactive efforts to discover, disclose, and correct environmental noncompliance will still result in the assessment of a fine. Moreover, culpable employees who commit crimes are specifically excluded from coverage. Businesses that already conduct routine audits or have implemented compliance programs need to adjust their procedures to account for new provisions of the policy. The policy's provisions must be read in conjunction with other government policies, and collateral consequences considered before a decision is made to disclose an environmental violation of law to the authorities.

During the public comment period, there were concerns about the EPA's draft interim policy and valid complaints regarding its limited scope and internal flaws. Much of that criticism was directed at the narrow and nonbinding wording of an interim policy, some of which was deleted from the final policy. The most controversial provision which remains, however, provides the EPA with broad authority to penalize a business for the economic benefit it gained before the violation was discovered, remediated, and disclosed. Simply put, the EPA is not willing to offer total immunity to good corporate citizens who, through diligent efforts, discover and correct environmental noncompliance.

According to the EPA, Florida and California have adopted policies similar to the EPA audit policy for violations of state environmental laws. Twenty other states have passed legislation that offers businesses a range of legal incentives up to and including immunity from civil and criminal penalties if they take measures to discover, correct, and voluntarily disclose such violations to state authorities. The EPA has openly criticized these states on the basis that they are unnecessarily protecting businesses from strict environmental enforcement. Senior officials at the EPA have called Virginia a leading example of widespread resistance by some states against vigorous enforcement of environmental laws (Cushman, 1997) . Virginia officials quickly responded that EPA has been heavy-handed and rigid for political reasons and is distorting Virginia's enlightened flexible approach to environmental protection.

On July 16, 1996 the National Governors' Association (NGA) issued an environmental self-audit privilege statement that declared that states should be able to institute voluntary environmental self-audit programs without interference from the federal government. The NGA demanded that the EPA defer to the states' decisions regarding the implementation of self-audit privilege statutes "unless it can be demonstrated that such state programs could significantly undermine state enforcement authority or otherwise obstruct the achievement of environmental objectives established by a federal program."

The conflict between the EPA and state legislatures has prompted a move by Congress to pass national audit privilege legislation that goes beyond the scope of many of the audit laws enacted by the 20 states. *Senate Bill 582*, introduced in 1996 by Senators Hatfield and Brown, provides complete immunity from prosecution for businesses that voluntarily report environmental violations and satisfy other requirements of the law. The EPA and a number of prosecutors opposed this legislation. According to the testimony of Steve Herman on May 21, 1996 before the Judiciary Subcommittee on Administrative Oversight and the Courts, this legislation "should be rejected for the special interest pleading which it is." Mark Woodall, Chairman of the Sierra Club's audit privilege task force, called audit privilege legislation in general "the worst idea of corporate thuggery I've ever seen." Industry sources say that the legislation will be redrafted in 1997 to address the EPA's concerns with the original bill.

Some critics, on behalf of the business community, have charged that the EPA rushed its audit policy without going through usual rule-making where there would be an opportunity for notice and comment on the proposed policy. They argue that the policy was enacted to stall legislative initiatives by many other states that were considering joining those states with already enacted immunity laws. Senator Hatfield said at the hearing on the 1996 proposed audit privilege legislation that the experience of those states should be gathered and distilled before Congress can determine what the appropriate level of federal legislation for environmental audits should be.

The EPA audit policy bears close inspection to determine whether it will encourage businesses to comply with the law. Businesses that choose to ignore the policy and fail to establish auditing or effective compliance programs are making a bad business decision. Unprotected businesses that commit violations of environmental law can expect to be prosecuted or penalized with civil fines by either, or both, federal and state governments, and likely will suffer the harshest consequences.

HOW THE EPA POLICY WORKS

The policy requires a business to exercise "due diligence" in responding to environmental noncompliance. "Due diligence" is defined as "the regulated entity's systematic efforts, approximate to the size and nature of its business, to prevent, detect and correct violations . . ." When a business exercises due diligence and discovers environmental noncompliance, it will receive the benefits of the policy if four conditions are met.

Audits and Internal Compliance Programs

First, the violation must have been discovered through either (1) an environmental audit, which is defined as "a systematic, documented, periodic and objective review by regulated entities of facility operations and practices related to meeting environmental regulations"; or (2) an objective, documented, systematic procedure or practice that reflects the regulated entity's due diligence in preventing, detecting, and correcting violations. The policy provides that the business "must provide accurate and complete documentation to the Agency as to how it exercises due diligence to prevent, detect and correct violations . . ." The Agency is providing incentives to businesses to create internal compliance programs and states that the criteria are flexible enough to accommodate different types and sizes of businesses. The EPA "recognizes that a variety of compliance management programs may develop under the due diligence criteria, and will use its review under this policy to determine whether basic criteria have been met."

Voluntary and Prompt Disclosure

Second, the violation must be identified "voluntarily," that is, not as a result of a government investigation or regular and authorized monitoring or sampling required by a statute. The policy specifically excludes emission violations detected through a continuous emissions monitor; violations of National Pollutant Discharge Elimination System discharge limits detected through required sampling or monitoring; and violations discovered through a compliance audit required to be performed by the terms of a consent order or settlement agreement. The policy will apply to any other violation that is voluntarily discovered, regardless of whether the violation is required to be reported. In response to comments pointing out that environmental reporting requirements are extensive, the EPA decided that excluding them from the policy's scope would severely limit the incentive for self-policing. When a violation occurs, the business must act quickly. The business has just ten days to disclose the violation in writing to the EPA. This period may be shorter where a statute or regulation requires reporting in less than ten days. Or, if reporting within ten days is not practical because "the violation is complex and compliance cannot be determined within that period, the Agency may accept later disclosures if the circumstances do not present a serious threat and the regulated entity meets its burden of showing that the additional time was needed to determine compliance status."

Prior to Government Investigation, Notice of Citizen's Suit or Other Process

Third, the business must identify and disclose the violation, and convince the EPA that it has done so prior to the following actions: the commencement of a federal, state, or local inspection, investigation, or information request; notice of a citizen's suit; legal complaint by a third party; the reporting of the violation by a "whistleblower" employee; or imminent discovery of the violation by a regulatory agency. This provision means that the business must have taken the initiative to find the violation and promptly report it "rather than reacting to knowledge of a pending enforcement action or third-party complaint."

Remediation Within 60 Days

Fourth, the business must correct the violation within 60 days, certify in writing that the violation has been corrected, and take whatever reasonable steps are necessary to remedy any environmental or human harm. The policy allows more time, if necessary, and the business may be required to agree in writing to take additional steps to prevent recurrence of the violation. Those steps include improvements to its internal operations or more systematic auditing. The business must also demonstrate "the specific violation, or closely related violation, has not occurred previously within the past three years at the same facility, or is not part of a pattern of federal, state or local violations of the facility's parent organization, if any, which have occurred within the past five years." At all times after the disclosure occurs, the business must cooperate "as required by EPA," which means it must provide at a minimum all requested, nonprivileged documents and access to employees and assistance in investigating the violation.

Penalty Reduction, Mitigation, and Criminal Referrals

If all four conditions are satisfied, then the EPA will not seek gravity-based penalties. Gravity-based penalties will also be waived "where the business can show that it has a compliance management program that meets the criteria of the policy." The EPA decided that it must recognize "voluntary efforts which play a critical role in protecting human health and the environment by identifying, correcting and ultimately preventing violations." If the violation is discovered by some means other than auditing, or as the result of a fully documented compliance program, the EPA will reduce any gravity-based penalty by 75% so long as all other conditions of the policy are met. The EPA expects that this reduction in the gravity-based penalty "will encourage businesses to come forward and work with the agency to resolve environmental problems and begin to develop an effective compliance management program."

The policy provides that the "EPA will not recommend criminal prosecution for [any business] that uncovers violations through environmental audits or due diligence, promptly discloses and expeditiously corrects those violations, and meets all [of the] other conditions ... of the policy." The EPA states that it has never recommended criminal prosecution of a business based upon a voluntary disclosure to the government before an investigation was underway. A review of the criminal docket reveals that there has never been a criminal prosecution for an environmental violation discovered as a result of an audit that was self-disclosed to the government. The policy is intended to reward "good actors." A criminal referral may occur, therefore, if the EPA determines that the violation is the result of "(i) a prevalent management philosophy or practice that concealed or condoned environmental violations; or (ii) high-level corporate officials' or managers' conscious involvement in, or willful blindness to, the violation."

THE RELATIONSHIP BETWEEN THE EPA POLICY AND STATE AUDIT IMMUNITY STATUTES

For several years, states have been enacting legislation that encourages businesses to conduct environmental audits and promptly correct noncompliance without fear of punishment from state authorities. Generally, these statutes treat an environmental audit as privileged as long as the business takes reasonable and prompt steps to correct the violation and notify the government of the noncompliance. A business that discovers an environmental violation, makes a prompt disclosure and takes corrective action will not suffer any penalty under these statutes.

The EPA is opposed to the establishment of a statutory environmental audit privilege because it believes that such a privilege "invites secrecy, instead of openness needed to build public trust in industry's ability to self-police." Business, however, is usually conducted without public inspection of confidential and, by their very nature, sensitive business records involving compliance issues. While the general public certainly has the "right-to-know" what types and quantities of hazardous substances are being used by businesses in their community pursuant to the provisions of the *Federal Emergency Planning and Community Right-to-Know Act of 1986,* presently there is no equivalent of the *Freedom of Information Act* for the public to obtain audit information directly from businesses. Businesses may volunteer to disclose this information to boost consumer confidence in their products or services, but there is no law or regulation that requires such confidential materials to be disclosed to "build public trust."

The EPA has expressed concern that the state audit privilege statutes are interfering with federally delegated programs. The EPA released a memorandum on February 14, 1997 which is intended to guide state officials whose states have enacted audit privilege statutes to avoid losing federally delegated programs. States run the risk of losing those programs in the event they do not have the ability to obtain immediate and complete injunctive relief against companies that self-report environmental noncompliance. The guidance also provides that states must retain the ability to collect civil fines for significant benefits derived from violations, repeat violations, violations of judicial or administrative orders, or serious harm and actions that pose an imminent and substantial danger to health or the environment. The guidance document may indicate that the EPA is beginning to change its policy, which up to this point has been overtly hostile to states with audit privilege statutes.

The EPA has argued in the past that there is no evidence that a privilege is needed. This is contradicted by the 20 state legislatures that passed environmental audit privilege legislation between 1993 and 1997. These states recognized that environmental regulations are comprehensive, complicated, and frequently violated by the regulated industry. In order to improve compliance, businesses need to conduct audits to identify noncompliance, to correct it as soon as possible, and to report it to the authorities to prevent recurrence.

Eleven other states and both houses of Congress recently considered enacting an audit privilege. In addition, the Texas Natural Resources Conservation Commission, the Maryland Department of the Environment, the Minnesota Pollution Control Agency, and the New York State Department of Environmental Conservation joined many businesses in arguing that businesses should be provided with greater incentives to audit their operations and to ferret out noncompliance without suffering harsh financial consequences from the government. The EPA should not ignore this sizable number of legislators, government regulators, and business people who believe that the state audit privilege is needed.

The EPA has argued that state legislation "breed[s] litigation as both parties struggle to determine what material [falls] within its scope." No such problem exists in Colorado, where 26 entities have already disclosed violations without spawning any litigation. A review of the Colorado Department of Public Health and the Environment (CDPHE) records shows who is disclosing violations and a brief description of the types of violations that are being reported:

Colorado Violations

Violations were reported by the City of Pueblo, the City of Fort Collins, the Denver Water Board, the Denver Metro Wastewater Reclamation District, several small companies and some large companies including Ball Corporation, Chemical Waste Management, Colorado Interstate Gas, the Gates Rubber Company, Phillips Petroleum, and Total Petroleum. The violations include:

- Discharge of groundwater contamination from a solid waste disposal facility
- Failure to file APENs and obtain air permits
- Discharges of wastewater with a slight oil sheen into a former cooling tower pit and a scrubber pit
- Discharges without a *Clean Water Act* permit
- Unpermitted discharges into sanitary sewers
- Noncompliance with water permit limits
- Storage of hazardous waste in violation of permit conditions
- Failure to maintain up-to-date APENs
- Other related environmental reporting violations

CDPHE has granted immunity from prosecution in 19 of these cases, denied immunity in 2 cases and has an open investigation pending in the remaining 5 cases involving the cities of Pueblo and Fort Collins, Western Mobile Northern, Inc., Colorado Interstate Gas, and the Denver Water Board.

Public interest groups opposed the Colorado audit privilege statute and have reportedly attacked CDPHE for being too soft on these violators. On January 29, 1997 Earthlaw; the Sierra Club; the Oil, Chemical, and Atomic Workers International Union; the Western Colorado Congress; and the High Country Citizens Alliance petitioned the EPA to stop delegating its authority to Colorado to provide permits and enforcement under the *Clean Water Act*. These groups argue that Colorado is no longer enforcing the delegated Federal programs and is not requiring compliance with environmental laws. A similar petition was filed in Ohio on the same day. These petitions have aggravated an already uncomfortable relationship between the states that have enacted audit privilege legislation and the EPA. Some believe that the petitions have strengthened the EPA's hand in its long-running dispute with these states.

The EPA has recently increased its enforcement efforts against businesses that have settled their environmental problems with state environmental protection agencies. For example, on March 18, 1997 the EPA charged Conoco with 78 violations of hazardous waste laws and assessed $666,771 in penalties against the business. This action occurred 11 days after the State of Colorado settled state charges on the same violations against the business by assessing $33,000 in fines. There was no violation of the Double Jeopardy Clause of the *U.S. Constitution* because the federal and state governments have the right to prosecute the same violations of environmental law using federal and state statutes. Conoco's spokesperson said the business was "astonished" by the EPA's "over-filing." In another case announced the same day, the EPA fined Platte Chemical of Greeley, CO $1.2 million for 752 violations of the federal *Resource Conservation and Recovery Act* (RCRA) where the state had been planning to assess about $400,000 in fines, according to the state's hazardous waste coordinator, Howard Roitman. In still another case, the EPA charged a Denver radiator and air-conditioning business with 257 hazardous waste violations and a proposed fine. The fine was $300,000 higher than a penalty that was included in a settlement with the state one day earlier. The EPA commented on these cases by saying that the state did not take timely or adequate action.

Political rhetoric has increased since Colorado passed its audit privilege law in 1994, particularly with inflammatory headlines in the *Denver Post* like "Polluters tell all, get off scott-free" (April 21, 1996). The Acting EPA Regional Administrator in Denver called the Colorado bill "the worst of the worst" because it had "all kinds of potential for abuse."

Texas Violations

Since Texas' audit privilege statute took effect on May 23, 1995, 239 entities have notified the Texas Natural Resource Conservation Commission (TNRCC) of their intent to conduct environmental audits. These entities include universities, the U.S. Air Force, municipalities, and many businesses, including petroleum and chemical companies. Already, 35 have disclosed reports of violations to the Commission. According to the author of the legislation, the TNRCC would never have discovered these violations if the state had not enacted the audit immunity statute. A review of the types of violations disclosed reveals the following:

- An audit of the Lyondell Petrochemical Business determined that a small percentage of the wastes burned in its East Plant boilers may be considered a hazardous waste.
- An audit of Southwest Shipyard, Inc. revealed 33 violations of air quality, hazardous waste, spill prevention control, storm water pollution and wastewater at its Channelview facility.
- An audit of Diamond Shamrock's Three Rivers Refinery found that three employees who routinely perform inspections under RCRA have not met annual training requirements.
- Enron Corp. disclosed that an incomplete audit of its Pasadena facility showed that some volatile organic compound vapors were not sufficiently controlled.
- Hurley Holdings Inc. stated that an audit of its Dallas facility demonstrated that it had "certain environmental problems associated with activities conducted at the site, which primarily resulted in surface and subsurface soil contamination involving certain solvent constituents."
- Hoechst Celanese Corp. disclosed that an audit of operating logs for its Pampa chemical plant last year found erroneous log entries. The responsible employee was immediately fired.
- Union Carbide Corp. stated that it had completed the action items recommended from two audits of its Seadrift Plant last year.

These early results of the Colorado and Texas statutes indicate that many businesses and other organizations, such as educational institutions, municipalities, and the military will take advantage of the requirements of the law by conducting audits, discovering violations, and taking corrective action which includes the discharge of responsible employees. Most of the violations appear to be the result of either careless employees or unintentional, but quickly corrected, mistakes. This experience demonstrates that the audit privilege is serving a useful purpose by having companies take positive steps to determine whether violations have occurred and forcing those responsible to take immediate corrective action.

ADVANTAGES OF THE EPA'S POLICY

The EPA's policy allows businesses to get credit for discovering and remediating noncompliance. Businesses may or may not have an environmental management system to effectively manage and reduce environmental risk. But by referencing compliance programs, the EPA is encouraging businesses to create procedures that are designed to identify and correct problems that cause environmental noncompliance and to eliminate the source of the problem before a violation occurs. Businesses with strong leadership and commitment to environmental excellence have already implemented compliance and risk reduction systems with the goal of minimizing serious environmental risk. These businesses are the most likely candidates for a 100% gravity-based penalty reduction if a violation should occur. The EPA's policy sends a strong message to businesses without environmental management systems that they should consider taking that action soon.

For businesses considering the adoption of an environmental management system, it is necessary that the procedures be fully documented. The program must include a records retention component designed to preserve evidence of compliance and demonstrate a commitment to environmental excellence. Chapter 16 sets forth how such a component can be incorporated into an environmental management system. If an infraction subsequently occurs, the documentary or electronic evidence needs to be marshalled and presented in a timely and convincing manner to demonstrate good corporate citizenship. Those businesses that ignore the provisions of the policy, or simply do not have the resources or the motivation to conduct audits or create an environmental management system, may receive up to a 75% penalty reduction for a disclosed violation. The violation must be discovered in the ordinary course of business, promptly disclosed, and expeditiously corrected. It is hard to justify the EPA's policy in treating these businesses differently where the violation is handled in this manner.

The policy states that the "EPA has never recommended criminal prosecution of a regulated entity based on voluntary disclosure of violations discovered through audits and disclosed to the government before an investigation was already underway." This should allay some fears in the regulated community that the EPA will misuse the policy as an artifice to lure unsuspecting businesses into conducting audits or creating compliance programs to increase enforcement actions. This clearly has not occurred in the first year and a half of the policy's existence. The EPA only needs to cite its record of criminal prosecutions (see Chapter 15) to demonstrate it does not need to use audits against businesses to initiate those prosecutions.

According to the EPA enforcement report issued on July 23, 1996, the EPA began a record 562 criminal investigations in FY 1995; 256 of those investigations have been referred to the U.S. Department of Justice for prosecution. The EPA has no need to violate its internal policy by subpoenaing audits from businesses to find evidence of criminal wrongdoing. Moreover, given the low threshold of criminal intent necessary for conviction of environmental crimes, an audit report, while useful, is not needed to convict a business of a criminal violation. In any event, many environmental criminal cases are commenced on the basis of disclosures by disgruntled employees to the authorities. Businesses that adopt and implement environmental management systems with effective auditing procedures can build in the means to address employee concerns, eliminate the causes of environmental noncompliance, and thereby further reduce environmental risk.

DISADVANTAGES OF THE EPA'S POLICY

The EPA has reserved the right to recover any economic benefit, which is not "insignificant," gained as a result of noncompliance. This recovery preserves a "level playing field" and prevents violators from obtaining a competitive advantage over other businesses that comply with the regulations. The EPA explains that an economic benefit penalty is needed. Otherwise, businesses have no incentive to comply on time, and responsible businesses will be undercut by their noncomplying competitors.

There is a rational basis for maintaining an economic benefit fine for intentional violations of environmental law. Such actions might place a wrongdoer in the position of competitive advantage over other companies that make significant capital expenditures and take steps to develop compliance and auditing programs to achieve regulatory compliance. Neither of the EPA's claims, however, justifies an economic benefit fine for unintentional environmental violations.

Timing is not an issue for these types of offenses because a business must disclose its actions within ten days of discovery of the violation. Proactive businesses that create environmental management systems to avoid committing violations, and promptly disclose and correct an unintentional violation as soon as it is discovered, should realize at most a marginal economic benefit from the noncompliance. The EPA policy suffers from imprecision because it does not distinguish between unintentional and deliberate violations. There is no definition of "significant economic benefit," although that does not appear to have hindered the EPA in negotiating settlements with companies that have disclosed environmental violations. Economic benefit could be defined in this context as the avoided cost of not having to comply with the law. Correcting the problem, by definition, eliminates this "benefit" to the business. There is no valid reason to further penalize a business for unintentional acts where it has engaged in immediate, good-faith efforts to detect violations and report them to the government.

Another compelling reason the EPA should consider deleting the economic benefit penalty for unintentional acts from the policy involves the general theory of setting standards. Thomas McBride, a professor of managerial accounting at the University of Massachusetts, describes the logic behind setting standards in this way:

> When any organization sets standards, there should be the realization that perfection is not an achievable goal. This is particularly evident with complex processes or virtually any activity involving human behavior. To set 'zero tolerance' standards ignores the natural occurrence of minor variances that are inherent in essentially all processes. Standards are established in the first place as a means of identifying abnormal activity or results so that corrective action can be taken to return that activity or process to normal. Since perfection is an

impossible goal, focusing on variances based on it can be a
waste of time and resources. Setting standards based on perfection can also result in the opposite reaction expected or
desired because even the most diligent efforts are doomed to
fail to meet the standard. A more practical, realistic approach
would be to establish standards that are rigorous or even demanding, but ultimately, achievable.

Some critics have charged that the EPA's policy does not provide sufficient
incentives for businesses to conduct comprehensive audits. They argue that fear
of disclosure is a disincentive. Audits that are not subject to privilege will be less
candid, making it more difficult for compliance officers to get funding from
corporate officers for environmental management systems. There are also some
concerns that the policy has no effect on clean-up costs, remedial costs, natural
resource damages, or emergency response costs. The U.S. Department of Justice is also not obligated to abide by this new policy and refrain from prosecution just because the EPA decides that it will not recommend prosecution.

Others argue that the agency will demand the production of audits conducted
under consent decrees or settlement agreements. There is nothing wrong, however, with the government seeking audits in these special circumstances. Consent decrees or settlement agreements by their very terms can require businesses
to conduct periodic audits to assure government officials that the business is
complying with the law and has taken responsible steps necessary to correct
previous problems and maintain compliance. The EPA has the right to review
the results of audits after violations have occurred and to monitor the performance of such businesses to determine whether they are abiding by the terms of
the consent decrees and settlement agreements.

The EPA will demand production of environmental audits when a business
puts its management practices at issue by pleading them as a defense or arguing
for penalty mitigation. But if a business is under criminal investigation and the
government obtains a prior audit by grand jury subpoena, that evidence may be
introduced at trial to prove the underlying substantive offense. It may also be
admissible pursuant to the *Federal Rules of Evidence* as impeachment evidence
of other crimes, wrongs, or bad acts. A regulated business has no immunity from
a federal criminal investigation because it has conducted an environmental audit. The EPA has not placed a limit on how far back in time it may go to search
in a business's records for incriminating evidence. This is no different from a
money laundering investigation, for example, where evidence of previous transactions from any branch of a financial institution can be subpoenaed for purposes of determining the scope of a criminal conspiracy or unearthing further
evidence of a pattern of wrongful acts.

It is true that employees are fair game under the provisions of the policy and
may be prosecuted, even if a business discloses and cooperates with the government. In one recent case, a parent corporation discovered that its subsidiary was
violating environmental laws. The parent turned in the subsidiary and its senior

management and thereby avoided penalties for itself under the EPA audit program. The subsidiary and its employees were charged with criminal offenses. Employees have never been immune from prosecution simply because the business for which they worked was cooperating with the government. Moreover, a policy that exempts employees who commit wrongful acts simply because the business conducts an environmental audit would be inviting wrongdoing or inattention to environmental regulatory responsibilities.

Critics argue that exempting employees does not make sense. It is doubtful that anyone in the business will make a voluntary report of a violation to authorities where that person could be criminally responsible for making the report. This analysis is simply wrong because the government would be foolhardy to prosecute an individual for the act of reporting the violation instead of the underlying wrongful act that is reported. There is a huge difference. An environmental management system that instills the responsibility at all levels of the organization to comply with the law and make prompt disclosures of unlawful conduct is the best way to assure compliance.

DECISION TO DISCLOSE

If a business discovers a regulatory violation during the course of an audit, the following are some considerations that need to be addressed before disclosing the violation to the government.

First, it will be necessary to correct the violation as soon as possible or to take reasonable steps to identify the source of the problem and create a corrective action plan. The business must always demonstrate good faith.

Second, the business must determine whether it has a legal duty to disclose the violation. Ten days is not a lot of time to make this decision, which must be based on an analysis of facts and law. For this reason, many businesses are creating environmental management systems with a mechanism in place to have responsible management employees have automatic authority to collect information and to handle decisions involving violations. The environmental management system can prevent violations, but it also lets operations personnel know immediately whom to contact, how to respond if a violation occurs, and what information is needed. Disclosure to the authorities is likely the preferred option as the government may not prosecute or may mitigate fines and civil penalties for a prompt voluntary disclosure.

Third, it will be necessary to get an objective opinion of knowledgeable environmental professionals, including an environmental consultant and an environmental attorney, to discuss how to take immediate corrective actions and whether to make a disclosure to regulatory authorities. Those professionals will also determine the range of consequences to the business and a comparative analysis of the punishment, if any, imposed upon similarly situated businesses that made voluntary disclosures.

Fourth, the business should consider all the related and unrelated collateral consequences that may occur as a result of a disclosure. For example, there may be third parties, like neighbors and environmental groups, who may have legitimate concerns and may be entitled to information about the violation. These individuals are not bound by the EPA's policy and may decide to bring their own citizens' suits against the business for damages to the environment.

Fifth, the business should consider how to eliminate the cause of the violation by implementing an environmental management system or changing the existing program to prevent recurrence of violations. One goal of such a system is to prevent and deter violations of law. A dynamic program will have procedures in place by which management can analyze and restructure the program when a violation has occurred in order to prevent future violations.

CONCLUSION

The EPA audit policy provides solid reasons for businesses to create environmental management systems with audit procedures to discover environmental violations and correct problems as expeditiously as possible. While the EPA should avoid imposing economic benefit penalties on good corporate citizens who have made the commitment to maintain compliance by designing effective programs, this should not prevent businesses from correcting violations and reporting them to the government.

Businesses need to include in their environmental management systems enough information regarding the EPA's audit policy to make an informed and timely decision that will avoid litigation over environmental violations. Chapter 15 addresses additional ways in which businesses can avoid committing environmental crimes. In the next chapter, we discuss various means of avoiding environmental civil litigation in general and cost reduction strategies in particular when such a dispute arises.

Chapter 12

REDUCING LITIGATION
RISKS AND COSTS

A corporate risk reduction strategy will be ineffective unless it has a mechanism to reduce the threat of environmental litigation. For any small- to mid-sized business, environmental litigation can drain a business's resources. Even large businesses that have gone the distance and tried environmental cases to completion sometimes refer to them as bet-the-company cases. Some environmental cases involve as many as 1000 parties and have created "monstrous practical problems for the federal courts ..." (Mininberg and Goodbody, 1994). A typical Superfund action may last five years or more from the time of filing until disposition in the district court. Lengthy appeals and even lengthier contribution actions against other potentially responsible parties (PRPs) often follow.

The best way to avoid litigation with the government or private parties is to be in compliance 100% of the time and to have no accidents. This is an extremely difficult task to accomplish for any size of company, but particularly for large companies that have far-reaching environmental impacts and responsibilities in multiple jurisdictions. Obviously, an effective environmental management system is the optimal means to avoid environmental disputes and thereby minimize environmental litigation risks.

The Corporate Counsel Section of the New York State Bar Association recently issued a report, together with a litigation management model, aimed at showing businesses how to avoid litigation and to reduce its attendant costs whenever it occurs (Haig, 1997). The report emphasizes the need to consult with counsel before a dispute arises to prevent litigation or, alternatively, to simplify protracted legal battles. Preventive law is still such a new concept that it is not taught at most law schools — although many law schools, lawyers and judges are beginning to understand how certain preventive measures can be taken to avoid disputes.

PREVENTIVE MEASURES

The concept of prevention needs to be written into environmental risk management and risk reduction systems. The multiple methods of preventing problems need to be set forth with an alternative dispute resolution (ADR) component in the system if prevention fails and a dispute arises. Managers and employees need to be taught in the first instance the many procedures that can be used to facilitate problem resolution. These can include interactive training to teach employees how to consider collaborative decision-making techniques in solving everyday problems before disputes arise, as well as using ADR techniques such as mediation or arbitration when problems mature into disputes. The ADR techniques are considered in Chapter 13. Each of these exercises can demonstrate multiple approaches to risk management, risk transfer, and risk reduction. Involving counsel or other expert consultants in training and role-playing exercises can improve corporate awareness of the range of possibilities that can be experienced by the business when violations occur.

Legal counsel can further assist companies with the development of litigation awareness programs to identify liability-causing conduct. Specifically targeted written materials can be distributed. Seminars and other types of interactive training can be conducted for management and employees on a variety of topics including employment law, environmental compliance, and unfair trade practices — to name a few of the most litigated areas of the law.

A litigation awareness program can demonstrate how to implement a legal and regulatory compliance program designed to reduce the risk of litigation being filed by an employee or third party against the business. Educational programs can also teach employees how to avoid creating smoking-gun and other types of documents that can be used against the business. This is not an obstruction of justice course, but rather a technical and legal session on teaching employees not only how to avoid breaking the law, but also how to avoid creating documents that make it look like they broke the law. When documents are created, document control procedures need to be carefully implemented so that the business can maintain its compliance record in an organized and retrievable manner and prevent the destruction of relevant documents that could be used affirmatively to prevent a claim from being filed. Chapter 16 shows how document control procedures can be created for this and other purposes. Counsel can also review documents, including agreements, to assure that they include litigation avoidance mechanisms, such as arbitration and mediation clauses, to prevent unnecessary litigation over commercial and other types of disputes that frequently arise despite the best intentions of the business and its employees.

COLLABORATIVE DECISION MAKING

Many environmental cases have been brought by citizens' groups. Those cases have been filed either against the government to ensure that it meets statutory or regulatory responsibilities, time tables, and objectives, or against companies to prevent or be compensated for environmental contamination resulting in natural resource damages. Litigation avoidance in this context involves the ability of government officials or business leaders to meet with citizen leaders to engage in collaborative decision making to resolve disagreements before they ripen into disputes. Local, regional, and national services are available to assist companies, communities, government agencies, environmental groups, and others to resolve problems before a claim arises.

The Colorado Center for Environmental Management (CCEM), a nonprofit organization based in Denver, for example, has formed the E-net, a group of professionals including scientists, facilitators, mediators, public participation consultants, risk assessors, public policy specialists, environmental engineers, and lawyers to provide a team approach to avoid environmental disputes or to resolve them before they mature into litigation. Collaborative decision making provides an open forum for all stakeholders to be involved proactively in a decision making process before positions are polarized or in anticipation of conflict. E-net's focus of collaborative decision making is on involving stakeholders in defining the problems and in finding mutually satisfying solutions. The process also includes conflict analysis, process planning, management, and facilitation. If this process fails or a dispute has already been filed in court, E-net uses ADR techniques to avoid protracted environmental litigation.

FINANCIAL IMPLICATIONS OF LITIGATION

When a dispute erupts that cannot be resolved without resorting to litigation, businesses should fully weigh the financial and human resource implications of long-term environmental litigation (Helmstetter, 1996). Lawyers' and experts' fees, client resources, and other litigation-related costs will be significant. In one case featured in the *Wall Street Journal*, Motorola spent $15.2 million on defense costs in the first three years of pretrial litigation. In other cases reported recently in the *Denver Post*, Dow Chemical and Rockwell International have spent over $20 million in discovery-related expenses for a property diminution case involving a DOE weapons facility in Golden, CO. Costs in three other cases involving DOE weapons facilities total nearly $50 million. *A Civil Action,* by Jonathan Harr, details the chilling account of a group of citizens who brought a toxic tort case against two major corporations resulting in huge costs and unsatisfactory results for all involved.

Facing expenses of this or lesser magnitude and uncertain results, even Fortune 500 companies need to incorporate into their environmental management systems mechanisms to reduce the risk of protracted litigation whenever possible. The mechanisms include monitoring litigation costs closely when litigation occurs and reviewing options and strategies continuously as the litigation progresses to trial. A settlement strategy must be developed as early as possible and should be reviewed periodically as more facts become known in the pretrial discovery process.

Trial can be even more expensive, frustrating and time-consuming. The Occidental Chemical case involving Love Canal is a classic example. Love Canal came to national attention in 1978 when residues of chemical water began leaking into basements of homes abutting the site. The government relocated over 1000 families. Occidental's predecessor, Hooker Chemical, was found strictly liable and the government pursued punitive damages against Occidental. After

10 years of pretrial proceedings, the trial began in October 1990; 166 witnesses had already been deposed. The liability trial on the punitive damages theory of the government lasted eight months. There were approximately 4000 exhibits and 81 witnesses (50 lay and 31 expert). Final arguments took place on February 12, 1992, and the judge ruled for Occidental over 2 years later. In another case profiled recently in the *Natural Resources and the Environment Journal*, the defendants successfully litigated the remedial phase of a CERCLA trial by using 45 consultants and experts (Coldiron and Bryan, 1996). Before a decision is made to try a bet-the-company case, or smaller environmental disputes, the business should consider ways to avoid the costs of protracted environmental litigation.

WAYS TO MINIMIZE LITIGATION EXPENSES

While the facts of every environmental dispute will be different, the basic approach to minimizing litigation expenses will be similar whether defending a government enforcement cost recovery action, a citizen's suit, or a private-party contribution action. To achieve realistic goals, a comprehensive litigation management plan should first be prepared. Inside counsel, or management if the business does not have inside counsel, needs to identify the company's overall goals and objectives for the litigation and create a preliminary cost-benefit analysis. These tasks should be accomplished before hiring outside counsel, if necessary, to handle the litigation (Haig, 1997). The Litigation Management Plan should include a discussion regarding concerns about adverse publicity, the negative impacts of litigation, as well as the benefits to be derived from the company having its day in court.

After the plan is developed, the business can either use its retained law firm or in-house counsel to handle the case, or it can consider soliciting bids and presentations from several law firms, and then select the best-qualified counsel to provide the desired legal services (Haig, 1997). A company can seek from applicants alternatives for handling the litigation, as well as a description of estimated costs for each strategy to meet the company's overall litigation objectives. All of this can be set forth in a privileged and confidential letter or memorandum which can further clarify the company's expectations from outside counsel.

Once counsel is assigned to the case, the litigation management plan can be expanded to incorporate his or her strategy as to how the case will be staffed and handled. Depending upon the type of case filed, the strategy can take into consideration the burden of proof on each element of the claim, the likelihood of injunctive relief being awarded, a liability assessment, imposition of fines or penalties, and a determination regarding recovery of response costs and allocation of those costs. The best way to reduce litigation expenses and meet expectations in a cost-effective manner is to prepare a comprehensive, privileged, and confidential litigation management plan and revise it periodically as circumstances demand. Settlement options should be discussed and weighed as the litigation proceeds towards trial. The following illustrates how to create an effective litigation management plan in the context of a cost recovery or contribution action filed under the *Comprehensive Environmental Response Compensation and Liability Act* (CERCLA) statute.

Reducing Litigation Costs Under CERCLA

When the government brings a CERCLA cost recovery action or a private party brings a contribution action, courts generally bifurcate the case into a liability phase and a damages phase. In the liability phase, the burden is on the plaintiff to prove a release or threatened release of a hazardous substance by a responsible party. The plaintiff must also prove that he or she has incurred costs. In a government cost recovery action, the burden then shifts to the defense to prove that the government's response costs have been incurred inconsistently with the National Contingency Plan (NCP). In a private-party contribution action, the burden remains with the plaintiff to demonstrate that its costs were incurred consistently with the NCP.

In the prelitigation setting, a business, like Occidental or Motorola, may have already become deeply involved in developing a defense or other risk management strategies. To be able to accurately predict the future course of litigation, not necessarily its outcome, a litigation management plan can be developed to quantify how much it will cost, how long it may take to try the case, and what are possible settlement options. The plan has to focus initially on the following question: If the business is found liable, what will its share of the costs be? This question may well arise long before the case is filed. For example, the EPA may send a CERCLA *Section 104(e)* letter to the business seeking information about the waste disposal at the site, and the business may subsequently learn what amounts of waste other businesses contributed to the same site.

As a general rule, joint and several liability is often imposed in CERCLA cases, but it is not mandatory. Businesses can reduce liability if they can prove that they only contributed a certain percentage of the hazardous waste at a site. An analysis will determine if it is possible to (1) reduce the total amount of liability to an allocable share, and (2) quantify how much liability exists. In this process, it will be important to determine the business's percentage of contribution of waste material to the site, the business's rank in relation to the possible contribution of other PRPs, and the relative toxicity of the waste contributed in relation to other PRP wastes. The answer to these questions may result in a decision to settle the case if the business is a small contributor, or even a major player. Thus, in determining how to effectively manage a CERCLA case, the first important question to answer is whether the business should litigate at all.

Next, the plaintiff's (government's) best case needs to be addressed on an element-by-element basis. An analysis should be prepared of the probability that a court will conclude that the business is liable to compensate the plaintiff for past and future response costs. A detailed summary of the plaintiff's expected proof needs to be prepared to determine the relative strength of its case.

That will raise the next important question: To what extent, if at all, should the business litigate the case? In determining how hard the plaintiff will press its case against this particular business, it must factor in what other businesses may be liable for and if they are able to pay an allocable share of the response costs.

For years courts have narrowly construed CERCLA's liability defenses (act of God, act of war, or act or omission of a third party) and have provided very few opportunities for businesses to escape liability. Recently, there have been a spate of important cases that have opened up new avenues for statutory defense theories to CERCLA cases. In one recent case, for example, the entire defense was predicated upon a mistake made in cleaning up an allegedly nonhazardous substance. Other defenses have focused entirely on whether a "disposal" occurred within the meaning of the statute or whether the site was a "facility." Each statutory defense must be reviewed to determine the likelihood that the court will accept it.

Before commencing the defense, in-house counsel and/or senior management must consider:

- What will be the opportunity, if any, to develop the defense during discovery (the government has statutory authority to limit discovery)?
- How much will it cost in lawyers' and experts' fees to prepare and present the defense?
- Will there even be an opportunity to present the defense in a liability trial?

Following this analysis, if the business's defense is not compelling, settlement options should again be reviewed and possibly pursued.

In CERCLA actions it is important to organize the codefendants into a single cooperative force (Helmstetter, 1996). Cocounsel from several firms can divide research and briefing tasks to ease the financial burden. A PRP agreement can be drafted which will present a united front to the government. This means the defendants will have to develop some method of alternative dispute resolution for internal allocation issues (Helmstetter, 1996). Codefendants can share the cost of experts and improve trial strategy by using their combined expertise in handling complex environmental litigation.

One possible technique to simplify trial issues is to stipulate liability if there are no valid defenses. This eliminates issues from the case that can cast the defendants in a bad light (Helmstetter, 1996). If the decision is made to contest liability and the court concludes that the defendants are liable, then the defendants must prove that the government's costs are inconsistent with the NCP. There have only been a handful of cases in the past 17 years where defendants prevailed in this type of proceeding, usually in the Court of Appeals. Far more frequently, courts defer to the EPA's expertise in choosing a suitable remedy and incurring costs in a manner consistent with the NCP. By contrast, private-party plaintiffs in contribution actions have to prove their costs were incurred consistently with the NCP. There have been many cases where plaintiffs have been unable to obtain contribution because they did not follow the NCP in conducting

"CERCLA-quality" clean-ups.

Valid reasons sometimes exist for challenging the extent of the government's remedial efforts in a cost recovery case. Experts can be retained to conduct allocation analyses and reviews of government documentation. It is not sufficient, however, to merely file expert affidavits challenging individual site assessment and clean-up costs as excessive and unreasonable. Defendants must have evidence that the government acted arbitrarily and capriciously in failing to consider cost or in selecting a remedial alternative that is not cost effective. Therefore, a candid assessment must be made as early as possible in the liability phase whether there is any chance of prevailing on these cost recovery issues. Most of the time, the best that can be done in the cost-recovery phase is to convince the government, a remote possibility, or a court that some of the expenditures are inappropriate, and thus not compensable. In some cases where the defendants' experts have made a showing of inconsistency with the NCP, the government has voluntarily withdrawn some response costs.

Insurance Coverage

As early as possible, an investigation must be undertaken to determine whether insurance policies provide coverage for environmental damage, as discussed in Chapter 8. The historic review should include:

- Comprehensive General Liability (CGL) policies that were purchased before 1970
- CGL insurance policies that were purchased between 1970 and 1985 which contain pollution exclusion clauses
- Post-1985 CGL insurance policies which contain absolute pollution exclusion clauses
- Personal liability insurance coverage (frequently sold with general liability policies and with excess and umbrella policies)
- First-party property damage insurance policies
- Other insurance policies including environmental impairment liability insurance (EIL)
- Insurance policies of predecessor organizations
- Other parties' insurance coverage (Anderson, Devries, and Rodriguez, 1994)

When this information is collected, an insurance profile can be prepared as part of the litigation strategy that identifies all insurers, coverage periods, exclusionary clauses, and notice provisions. Litigation risks are reduced if satisfactory insurance coverage can be located to indemnify the business for the costs of defense and damages. Timely and adequate notice to these insurance carriers must be given. Insurance agents are also a valuable source of information and can be used to notify carriers of potential claims. If policies are missing, then a secondary search for evidence of coverage is necessary which would include invoices, cancelled checks, policy renewals, and correspondence. Insurance archaeologists who specialize in finding copies or evidence of lost policies can be helpful. The existence of comprehensive general liability policies will be significant if it can be proven that they indemnified or partially indemnified the business for environmental liabilities. When the policy profile is created, an assessment should be made as part of the risk reduction strategy whether additional coverage is required to eliminate or transfer environmental liability.

Locating Other Potentially Responsible Parties

At the same time, it is also necessary for the PRP group, or an individual defendant, to conduct a thorough investigation to locate other PRPs. There are several national PRP search firms that have considerable experience in conducting investigations of other companies responsible for past disposal of wastes. Recently, courts have been broadening the concept of owner/operator liability, which allows a business to include in contribution actions other businesses that stand in the chain of corporate succession. Additionally, courts also have been narrowly construing some indemnification provisions of contracts that preexisted CERCLA. Unless an indemnity provision specifically contemplated environmental liabilities or was so broad that it can be interpreted to have included those liabilities, it will not afford protection from liability. The more parties there are to share liability, the more money the business can save in the process. An important factor to consider is the cost of conducting a full-scale investigation, which may result in a contribution action against the other PRPs. As always, the likelihood of success on the merits needs to be balanced with the cost of proceeding against other PRPs for contribution.

CONCLUSION

To avoid the high cost of allocation litigation, consideration, either individually or as a group, should be given to hiring an allocation consultant. This could lead to a negotiated settlement, sometimes with the help of ADR, discussed in the next chapter. Well over 90% of environmental cases settle before they go to trial. It is impossible to settle these cases unless counsel has sufficient knowledge of the facts to be able to give the client-company competent advice regarding the strengths and weaknesses of the opposing party's case. Unfortunately, the discovery process in environmental cases is slow and cumbersome, and it takes time and money to gain the requisite knowledge to provide useful advice to the client. Any efforts made by the parties to reach consensus regarding the facts and the identification and possible resolution of legal issues will speed up the process of litigation and ultimately save financial and human resources. Businesses need to consider alternatives to litigation during this process in order to adequately address all options in minimizing environmental litigation risk.

Chapter 13

SOLVING ENVIRONMENTAL PROBLEMS USING ALTERNATIVE DISPUTE RESOLUTION TECHNIQUES

As explained in the previous chapter, environmental litigation is deleterious to business interests and can be and should be avoided whenever possible. In sophisticated and complex environmental cases, lawyers', consultants', and experts' fees mount quickly. In addition to the expenses involved, backlogs and other delays frequently occur so that resolution of cases takes many months and, in Motorola's, Rockwell's, and Dow's cases, years. Litigation is always disruptive to business and outcomes are never certain. Settlement in most civil cases occurs after discovery is completed and a trial date is quickly approaching. By that time, the parties have become so polarized that collaborative outcomes are unlikely and relationships, to the extent that they exist, are strained, if not broken forever.

In environmental cases, there are additional compelling reasons to avoid protracted litigation. The judiciary, as good as it is, lacks technical training and oftentimes has difficulty adjudicating the complex scientific evidence that is frequently the main subject matter of environmental disputes. Recently, the Federal Judicial Center recognized this problem and published the *Reference Manual on Scientific Evidence* to assist judges in managing expert evidence, primarily in cases involving issues of science and technology. When the government brings an environmental enforcement or civil action, it is rare that a business prevails in the trial court. Most victories for businesses have occurred in the courts of appeals after lengthy and contentious trials and expensive appellate proceedings. Many business leaders have understandably become intolerant of these uncertainties, in part because litigation is controlled by judges and lawyers. Others are frustrated with the transactional costs and delays in solving problems and disputes. While the courts decide legal issues, they do not and cannot resolve all disputes, nor can they assure in all cases that parties are justly compensated for cognizable legal injuries (Dauer, 1994). Frequently, a successful defense can deplete corporate financial and human resources, and leave managers jaded by the experience.

Litigation often polarizes the parties as each side is encouraged to argue its position in a self-serving manner, emphasizing the degree of injury or lack thereof (Dauer, 1994). In these circumstances, collaborative outcomes are less likely. Litigation results in public scrutiny, which can have serious adverse financial

137

and public relations consequences. These disruptions divert management's attention from corporate goals and objectives. For all these reasons, lawyers and business people are seeking new and more effective ways to resolve disputes.

USE OF ALTERNATIVE DISPUTE RESOLUTION

In 1990 Congress passed the *Administrative Dispute Resolution Act*, requiring each federal agency to adopt a policy that addresses the use of alternative means of dispute resolution and case management. On October 23, 1990 President Bush signed an Executive Order requiring the use of alternative dispute resolution (ADR) as a tool to enhance the negotiating process and to avoid the disadvantages to all parties of protracted litigation. The year before the Act was passed, 220,000 civil suits were filed in U.S. courts, of which the federal government was a party in 55,000 of these cases. On February 5, 1996 President Clinton signed an Executive Order directing federal agencies to implement several strategies, including the use of ADR, to resolve civil claims by or against the government.

ADR is becoming a preferred method to assist parties in resolving disputes quickly and efficiently without the need for court intervention. In Colorado, over 600 large corporations have adopted the ADR policy statement that provides:

> In the event of a business dispute between our company and another company which has made or will make a similar statement, we are prepared to explore with that party resolution of the dispute through negotiation or ADR techniques before pursuing full-scale litigation. If either party believes that the dispute is not suitable for ADR techniques, or if such techniques do not produce results satisfactory to the disputants, either party may proceed with litigation.
>
>   (Center for Public Resources Policy Statement)

Different forms of ADR include early neutral evaluation, arbitration, mini and summary jury trials, and mediation. Early neutral evaluation involves either a judge, magistrate, or an outside person who provides a nonbinding assessment of the case. Arbitration involves an arbitrator, or a panel of arbitrators, who listen to presentations by both sides and then render a judgment on the law and facts. A mini trial is an abbreviated presentation of the facts to a mediator who may be joined by high-level principals from each side. The process is known as a summary jury trial when the presentation is made before a mock jury that renders a nonbinding verdict (Schiffer and Juni, 1996) .

As many environmental cases have been resolved using a mediator, this will be the primary ADR technique discussed in this chapter. Mediation uses a third party, with no decision-making authority, to assist disputants to reach a voluntary negotiated settlement. Mediators are usually former judges available through mediation services, local lawyers who mediate from time to time as an adjunct to their practices, and magistrates and presiding judges (Schotland, 1995). Mediation is a voluntary and confidential settlement procedure in which either party can back out at any time. Both sides have to voluntarily agree upon a mediator who assists the parties in reaching consensus without incurring the runaway costs of litigating the case to a verdict. Early resolution usually achieves the best financial results for all those involved. For businesses this means scarce resources will be preserved, and employees' time will be spent on more useful activities. There is also less likelihood that the corporate image will be tarnished in the marketplace and global community. The three most common advantages of mediation are that it helps bridge gaps between huge demands and low offers; it affords each party an opportunity to have a realistic appraisal of the strengths and weaknesses of its case by a neutral, detached and experienced third party; and it defuses personality conflicts that so often arise in the context of litigation and interfere with settlement prospects (Schotland, 1995).

For all these reasons, mediation works well in resolving civil cases. In the Florida state court system, it has been reported that between 60 to 70% of all civil cases settle at or shortly after mediation (Schotland, 1995). At Chevron, ADR-based mediation of one dispute cost $25,000, whereas going to court would have cost as much as $2.5 million (Carver and Vondra, 1994). Toyota set up a Reversal Arbitration Board to resolve disputes between the company and its dealers concerning the allocation of its cars and sales credits to the dealers. This new ADR procedure reduced the number of cases from 178 in 1985 to 3 in 1992 (Carver and Vondra, 1994). Using ADR techniques, AT&T Global Information Systems dropped the number of filed lawsuits from 263 in March of 1984 to 28 in November of 1993 (Carver and Vondra, 1994).

The EPA has wisely been advocating the use of ADR to resolve environmental disputes. In May 1995 the EPA reported that it has employed this technique to assist in the resolution of over 50 enforcement-related disputes. These cases range from two-party *Clean Water Act* actions to Superfund cases involving up to 1200 parties. Participants in an ADR pilot for Superfund cases reported the following benefits: "constructive working relationships were developed; obstacles to agreement and the reasons therefor were quickly identified; costs of preparing a case for U.S. Department of Justice referral were eliminated; and ongoing relationships were preserved" (EPA, Oct. 1991). The EPA insists that ADR techniques, including mediation, provide other benefits that will minimize transactional costs in resolving disputes.

The U.S. Department of Justice has reported that it has identified approximately 250 cases in which an ADR process, typically mediation, either has been used or has been considered as a mechanism for facilitating settlement. Some examples include:

- *U.S. v. Martin*: A CERCLA cost recovery case where a judge, appointed as a mediator, conducted a one-day mediation that required each party to present its case in a concise fashion. The judge and the parties then met in private sessions until a successful allocation agreement was reached.

- *In re Mountain Water Rights Adjudication*: A local rancher mediated a case involving a grazing association and the U.S. Fish and Wildlife Service. The rancher held a series of meetings with the parties, other stakeholders, state and local agencies, and sporting groups to conclude a successful agreement.
- *U.S. v. Farwest Fisheries*: A mediator was used to work with the parties to attempt to reach a settlement of an asbestos penalty case.
- *National Oilseed Processors Ass'n v. EPA*: A Court of Appeals mediator successfully identified a way for the parties to resolve their dispute in the appellate court.
- *U.S. v. 2.1 Acres*: A Court of Appeals mediator also assisted the parties of this case to reach a compromise (Schiffer and Juni, 1996).

Experience in these cases has shown that mediated negotiations encourage settlement as they tend to focus more on resolving the real issues separating the parties. Negotiations take less time and energy and are less expensive than litigation. The accuracy of the result of litigation is often dependent on the quality and the zeal of adversarial presentations (Dauer, 1994). Mediation is a preferable process in many cases because the parties can reach agreement more quickly by identifying common interests and tailoring them into settlement options to meet their specific needs. Instead of the adversarial process of litigation, mediation allows the parties to reach their settlement objectives efficiently, to minimize the damage to their ongoing relationship, and to alleviate cost and delay.

In the environmental arena, mediation is particularly appropriate to avoid costs, delays, and the uncertainty regarding ultimate results. The following are considered by the U.S. Department of Justice as reasons why ADR may be helpful in resolving environmental disputes:

- **The ability of a mediator to conduct frank, private discussions may improve the outcome.** For example, some litigants may be loath to "put the cards on the table" in front of other litigants. A mediator may be able to convey information to other parties in an indirect fashion.

- **The range of issues is broad enough, or can be creatively made broad enough, to allow tradeoffs and creative generation of options, especially when some options cannot be ordered by a court.** For example, in a *National Environmental Policy Act* dispute, underlying resource management decisions are likely the crux of concern, but cannot be reached by a court. By addressing concerns regarding the underlying dispute, a mediator may be able to fashion a resolution of the issue at hand. This result may forestall future litigation.

- **A mediator may be helpful in facilitating negotiation by breaking through impasses.** Such impasses may develop because of conflicts within interest groups, technical complexity, or uncertainty or political visibility or poor communication among the participants due to personalities or past history. For example, a mediator can defuse tension with a citizens' group concerned about a particular agency project by presenting negotiating proposals from all sides in an even-handed manner. In a case in which an impasse involves technical complexity, a mediator or a joint expert might even offer technical expertise on a given issue.

- **A thorough exchange of information may improve the outcome.** For example, a mediator can help to ensure that all issues are addressed, and that the heat of negotiating has not caused the parties to overlook an item that may be crucial to settlement implementation.

- **The participation of parties not directly involved in a legal action is necessary or beneficial to the settlement.** For example, numerous citizens' groups may be interested in a particular agency project; addressing the concerns only of the group that sued may be shortsighted, and invite future litigation from others.

- **The parties and issues are numerous, such that a facilitated, structured settlement process would be helpful, and no party is willing or able to take the lead role in establishing such a process.** For example, CERCLA allocation disputes often involve multiple parties and issues. A mediator who provides a structure for allocation can assist the parties in reaching a global settlement (Schiffer and Juni, 1996).

Certainly, businesses that are interested in resolving environmental disputes in a more efficient and cost-effective manner should consider mediation as a device to eliminate needless expenses and to maximize opportunities for a permanent solution.

REASONS FOR FOREGOING THE USE OF ADR

Why then do so many businesses and their counsel frequently forego using mediation to resolve environmental conflicts? There are many reasons. ADR is a relatively new concept in the environmental field. *Black's Law Dictionary* did not even contain a definition of ADR until its 1990 edition. It will take time before it is accepted by the judiciary, leaders of the bar, and the public as a worthy alternative to the legal process which has been with us for centuries.

In civil cases, ADR in general and mediation in particular will not work if senior corporate managers have a philosophy that winning is the only thing that matters (Carver and Vondra, 1994). Chevron, Toyota, and AT&T Global Information Systems successfully integrated ADR techniques into their business operations because their managers adopted the ADR principles wholeheartedly. As long as companies view ADR as the alternative, rather than the primary or preferred method of settling disputes, companies will not have obtained sufficient experience with it to make significant and necessary changes in how they litigate cases (Carver and Vondra, 1994). ADR is sometimes handled in the same way as litigation when the opponents "waste prodigious quantities of time, money and energy by reverting almost automatically to the habits of litigation" (Carver and Vondra, 1994). For these and other reasons, ADR has not been embraced as it should be by major companies throughout the U.S.

Many people simply trust judges to make the right decision rather than rely on mediators to bring parties to a negotiated agreement. Environmental cases are most frequently decided by judges and rarely by jurors. CEOs and senior managers trust judges will be fair to their cause and not overly penalize their businesses for unintentional environmental mistakes. Litigation also buys time. It allows senior managers to take the opportunity to make better reasoned decisions involving the business's resources. One environmental lawyer recently wrote that the passage of time can result in a potentially fairer liability scheme, more lenient clean-up standards, more reasonable agency personnel, and perhaps even improved site conditions through natural attenuation of the contamination.(Dean, 1996). Sometimes, if the environmental conditions are stable and there are no adverse health consequences involved, there may be no reason to make a decision until all of the facts and circumstances are known. A premature decision to resolve a dispute without full knowledge of the consequences can be severely prejudicial to all the parties involved. In some cases, corporate managers need to maintain control over the environmental problem, the nature and timing of the clean up, and the disclosure of contamination to the government so they can make correct decisions involving the financing and timing of remediation of environmental contamination. For these reasons they may not wish to agree to an expedited settlement using a mediator before there is an adequate factual basis to make a reasoned decision.

Mediation is best suited for situations where there is a reasonable likelihood of compromise. This does not mean that the parties think there is a reasonable likelihood of compromise, but rather the circumstances may allow such compromise. Parties in the throes of litigation may not be able to accurately assess each other's interests in early settlement without assistance from a neutral party. The chances of success increase significantly as the parties learn to trust each other and identify mutual interests that can be achieved through a negotiated settlement. Mediators develop special skills to encourage this to happen. The EPA has had success using mediation in cases involving a large number of businesses with liabilities that must eventually agree how to allocate their share of clean-up costs to settle their dispute with the government.

Another factor influencing the corporate decision whether and when to settle with the government is the uncertainty that occurs in most environmental cases. Both the state environmental enforcement agencies and the EPA have an extraordinary amount of regulatory authority concentrated in a few individuals who have the ability to charge companies and individuals with environmental violations. This starts a process that can inflict enormous burdens and financial damages on both individuals and businesses. The EPA, the state environmental protection agencies, environmental groups, and business leaders have been wrestling with conflicting theories of what works the best to clean up the environment, to minimize future impacts of business, and to deter bad actors from noncompliance. Is it more effective to increase enforcement of the federal, state, or local environmental laws, or to provide businesses with greater incentives to employ environmental management systems to prevent environmental violations and ensure swift clean up when noncompliance occurs? In the face of this ongoing controversy, many businesses are unable to tell when they should use alternatives to litigation to resolve disputes. When a business's continuing existence is threatened, it may have no other choice than to use all available resources to defend itself where liability is so sweeping and the future is entirely uncertain.

DECIDING WHETHER TO USE ADR OR LITIGATION

In determining whether to proceed with litigation or mediation after a dispute arises, it is important to weigh the merits of both objectively. Litigation and mediation are not necessarily mutually exclusive. There are circumstances in which multiple strategies can be developed using mediation at a later stage of the proceedings. This may occur after discovery of pertinent facts or by using a summary jury trial procedure to assess the strengths and weaknesses of a case to determine whether a favorable settlement is possible.

Factors that generally favor litigation include (1) when there is a need for a binding, enforceable and final judicial decision; (2) when procedural safeguards are necessary to improve the truth-seeking nature of the process; (3) when a fact-finding process is critical to resolution; (4) when an appeal system is necessary in the event of an adverse decision; (5) when established norms (i.e., principles of law and case law) are important to the outcome; and (6) when the outcome will further corporate goals, resolve uncertainties, or establish the rights and obligations of parties as a precondition to further legal proceedings (Dauer, 1994).

In determining whether to proceed with ADR, it is important to analyze future transactional costs for planning purposes. Accountants and management consultants are able to analyze these transactional expenses, including legal and consultant fees, to determine how much each stage of the proceeding should cost and whether other alternatives to litigation could be more cost effective for future scenarios. Each situation is different and requires independent assessment of various techniques that can be used to resolve problems expeditiously and cost effectively.

A recent case involving the Rocky Flats nuclear weapons factory near Denver, CO offers an opportunity to use ADR to cut down legal fees and costs in resolving a complex environmental dispute. Former Rocky Flats operators Dow Chemical and Rockwell International are defending a toxic tort class-action suit brought by 40,000 neighbors of the facility who allege that health threats have decreased their property values. Legal titans on both sides, including one plaintiff's lawyer from Cincinnati known as the "Master-of-Disaster," are squaring off for a classic legal battle to be waged in and out of court for several years before a jury is ultimately asked to reach a verdict. The case raises important public policy issues about how to resolve complex environmental disputes, particularly where the government is paying all defense costs because of indemnification agreements it had with the ex-operators.

How might mediation help resolve the Rocky Flats example? Both sides need to realize that jury verdicts, in most cases but particularly toxic torts cases, are wildly unpredictable. What is absolutely certain is the incurrence of extraordinary financial costs associated with bringing such a massive case to trial. For all parties, expert fees can range in the millions of dollars, even if the case does not go to trial.

The U.S. Supreme Court has recently made it more difficult to introduce expert testimony in cases where plaintiffs are pushing novel legal theories. In the Rocky Flats case, even if the testimony is admitted, jurors may not be swayed by dire predictions of health hazards when they find out, as has been alleged by the defendants, that the total release of radioactive emissions at the plant amounts to scarcely the equivalent of one chest X-ray per plaintiff. On the other hand, the defense will also be extraordinarily expensive; $20 million have been spent so far on handling preliminary discovery. All parties would benefit if they would consider a mediator or team of mediators to reach a consensus of this case without incurring further expense.

Legal-risk reduction requires planning and cooperation from all involved. A consensus-based approach to plan how to handle disputes when they arise serves to reduce financial and human resource impacts on the business. It is important for managers to understand that there are innovative techniques to be used to resolve disputes without incurring unnecessary costs. It is equally important for government officials to understand that it takes a cooperative effort between business and government to reduce environmental risks. The next chapter describes the EPA's initiatives in partnering with businesses that have had positive results of reducing environmental risks.

Chapter 14

VOLUNTARY PROGRAMS AND OTHER INITIATIVES TO REDUCE ENVIRONMENTAL RISK

In 1994 in its Five Year Strategic Plan, the EPA identified the need to develop and implement more innovative, effective, and efficient approaches to environmental protection (EPA, 1997a). The EPA is beginning to recognize that command-and-control regulations to achieve environmental improvements are only part of a multifaceted solution to environmental protection (EPA, 1997a). Accordingly, in the past several years the EPA has developed some innovative programs that are designed to encourage the government to work with the regulated community to minimize environmental impacts, reduce risks, and avoid environmental liabilities. There are a number of advantages for each business that participates in these programs, including recognition, streamlined regulatory permitting, and a reduction in the cost of doing business. Each of the programs is designed for partnerships to be developed among the EPA, the states, the businesses, and community stakeholders to find new innovative ways to reduce regulatory burdens and positively impact the environment. All of these programs involve or relate to the reduction of corporate environmental risk. The following is a description of the most well-developed voluntary programs and then a summary of an audit report which addresses whether two such programs have been effective in reducing environmental risks.

ENVIRONMENTAL LEADERSHIP PROGRAM

In April 1995 the EPA began a one-year pilot program which served as the basis for the creation of the Environmental Leadership Program (ELP). The goals of the program are:

- Better protection of the environment and human health by promoting a systematic approach to managing environmental issues and by encouraging environmental enhancement activities, such as biodiversity and energy conservation
- Increased identification and timely resolution of environmental compliance issues by ELP participants
- Multiplying the compliance assistance efforts by including industry as mentors

145

- Fostering constructive and open relationships among agencies, the regulated community, and the public

In order to qualify for this program, a business or federal facility must have a mature environmental management system, as well as a compliance and auditing program. That means that the environmental management system has gone through an initial development period. The business must participate in community outreach and employee involvement programs which foster relationships among facilities, employees, and local communities. Then, the business must submit to the EPA facility-wide compliance audit results and environmental management system information obtained within the previous two years. Federal facilities are required to verify that their parent agencies have endorsed the *Code of Environmental Management Principles* (CEMP) and to describe how the applying facility is implementing that program.

When a business or federal facility becomes a member of the ELP, the program is intended to facilitate an exchange of information. ELP encourages the implementation of best practices related to environmental management systems and pollution prevention activities. Businesses and federal facilities that successfully participate in the program will receive:

- **Public recognition:** the EPA will issue certificates of membership in the ELP and will develop programs and activities designed to publicly recognize ELP members.
- **Logo usage:** members can use the EPA-issued logo in facilities and advertising but not on products.
- **Inspection discretion:** the EPA may reduce or modify discretionary inspections.
- **Reduced regulatory burdens:** the EPA may consider in the future expediting members' permits, providing longer permit cycles, and streamlining permit modifications.

The EPA is encouraging businesses that get involved in the ELP program to develop environmental management systems within the framework of ISO 14001. According to the EPA, the purpose of ISO 14001 is to provide businesses with the elements of an effective environmental management system that can be integrated with other management requirements to help businesses achieve environmental leadership and economic goals. Becoming a member of ELP will not create a conflict with facilities seeking to become certified under ISO, nor will it require certification under ISO. The EPA is recognizing that ISO 14000 can be a useful management tool which, when integrated into the regulatory system, may allow businesses to obtain comprehensive, environmental, economic, and other benefits.

ELP facilities will be required to conduct compliance and environmental management system audits, either separately or combined. The results of the audits will be provided to the EPA and published in accordance with *ELP Audit Guidelines*. The results will be published in an *ELP Annual Report*, which will include information on the facility and its environmental impacts; the environmental management system's goals, objectives, and targets; and the audit results and the measurement of accomplishments and instances of noncompliance, if any. Members of ELP will have 60 days to correct noncompliance, or sooner if required by the EPA.

The EPA is anticipating that it will partner with states in developing the program by reviewing applications, participating in on-site reviews, selecting facilities, and implementing the program. Several states are in the process of developing their own leadership programs, and the EPA is developing a model EPA-State Agreement to facilitate these new arrangements. The agency is encouraging states to coordinate their efforts with the EPA to provide the greatest benefits to qualified facilities. The EPA is hoping the states will minimize duplication of effort and confusion resulting from multiple similar programs, while conserving limited regulatory resources. Just as in the case of the EPA's audit policy and the states' audit privilege statutes, the EPA is attempting to set national environmental policy and continue its supremacy of environmental regulatory authority. Putting aside the continuing power struggle between the EPA and the states, the ELP is an innovative approach to working with the regulated community in finding ways to reduce environmental impacts and to provide appropriate rewards for going beyond compliance with the law and regulations. The EPA's Project XL is a closely related program that contains nearly identical interests and goals for both government and business.

PROJECT XL

On March 16, 1995 President Clinton created Project XL (eXcellence and Leadership) as part of a 25-point program within his Administration entitled *Reinventing Environmental Regulation.* Project XL was originally designed to "give a limited number of regulated entities an opportunity to demonstrate excellence and leadership" (EPA Notice). The XL project was originally conceived outside of government by a multiparticipant process convened by the Aspen Institute (GEMI, 1996). The White House adopted the Aspen Institute's findings and placed the XL concept in a White House Policy on *Reinventing Environmental Regulation.* The proponents of the XL program decided to provide regulated industries, governments, and communities with flexibility to develop alternative strategies that will replace or modify specific regulatory requirements on the condition that they produce greater environmental benefits. When the program was initiated, the EPA stated, however, that "[i]n exchange for greater flexibility, regulated entities will be held to a higher degree of accountability for demonstrating project results."

Project XL has been controversial from the outset because of its heavy emphasis on accountability and its lack of a uniform approach to assessing environmental benefits and determining how they should be measured. Indeed, the EPA, public interest groups, state and local governments, and business leaders have been trying to reach consensus on what constitutes superior performance. How this issue is ultimately resolved may well determine how the EPA will reassess and redesign its environmental regulatory program as it modifies its regulatory focus from command-and-control regulations to more flexible regulations designed to provide positive incentives to business to take preventive measures to protect the environment.

On May 23, 1995 the EPA announced its intention to solicit proposals from the public and private sector for Project XL. The EPA originally placed a goal of 50 projects to be considered as part of the Project XL pilot program. The program's goal is to get a core group of companies and municipalities or other organizations to reduce costs and to achieve superior environmental performance beyond that required in existing regulations.

In November 1996 the EPA released draft guidelines for public involvement in the review of proposed XL projects. This addressed criticism of the program that the XL process set no parameters on participation (GEMI, 1996). In its announcement the EPA stressed the need for public involvement in the application-approval process. Interested parties are divided into three categories: direct participants, commentators, and the general public. Participants in the process have complained that the EPA's process already has become unwieldy. Community activists, with little technical expertise, have been negotiating with technical staffs of large businesses. The EPA responded by pledging up to $25,000 per project to provide public interest groups with resources, including technical assistance, to work with industry on XL projects. Some community activists have complained that businesses are being allowed to skirt environmental regulations. This criticism has dampened public and private interest in the program.

EPA's Selection Criteria

As of February 4, 1997 only 43 proposals have been submitted to the EPA for review, and almost half have been withdrawn. Ten are still in the development stage, three are in implementation, and one was successfully facilitated (IBM). Despite this shaky start, the EPA is convinced that XL will be successful and is refining its selection criteria to evaluate more XL proposals.

Projects will be chosen that are able to achieve environmental performance that is superior to what would be achieved through compliance with current and reasonably anticipated future regulation. According to the EPA, "'Cleaner results can be achieved directly through the environmental performance of the project or through the reinvestment of the cost savings from the project in activities that produce greater environmental results" (EPA, 1997b).

The EPA will review a proposal to determine whether it would produce cost savings or economic opportunity, or result in a decrease in paperwork burden. Project XL proposals must have significant stakeholder support. The support may come from stakeholders, including communities near the proposed project, local and state governments, businesses, environmental and other public interest groups, or other similar entities.

The EPA's selection criteria also focuses on projects that employ innovative strategies for achieving environmental results. One such proposal nearing completion is a site-specific, five-year permit tailored for Merck & Co.'s drug manufacturing plant in Stonewall, VA. Merck is planning to replace its coal-fired power plant with one that operates on natural gas, which will result in a 25% reduction in sulfur dioxide and a 10% reduction in nitrogen oxides. Merck is seeking a special permit that will allow it to make changes in its operation and equipment without undergoing lengthy reviews typically associated with plant modification. Merck has devised a three-tiered monitoring system designed to ensure compliance with its permit (EPA, 1997b).

The EPA's preference is for projects that have processes, technologies, or management practices that prevent the generation of pollution rather than controlling pollution once it has been created. The EPA wants projects that embody a systematic approach to environmental protection that tests alternatives to several regulatory requirements or affects more than one environmental medium. The EPA also prefers projects that are intended to test new approaches that could conceivably be incorporated into the agency's programs or in other industries or facilities in the same industry. The EPA is most interested in pilot projects that test new approaches that could one day be applied more broadly. Project proponents must demonstrate that they have the financial capability to carry out the project and that it is technically and administratively feasible (EPA, 1997).

Project proponents must also identify how to make information about the project, including performance data, available to stakeholders in an easily understandable form. Project proponents must state their objectives clearly so that requirements can be easily measured, allowing the EPA and the public to evaluate the success of the project and enforce its terms. Finally, each approved project must be consistent with Executive Order 12898 on environmental justice. That is, it must protect worker safety and ensure that no one is subjected to unjust or disproportionate environmental impacts (EPA, 1997).

The EPA provided three types of pilot project examples to help stimulate creativity. The examples were intended to be illustrative only of the various alternative strategies and new forms of flexibility that the EPA would entertain as part of its reinvention of government. In emphasizing its own flexibility in reviewing the proposals, the EPA stated that it encouraged "the submission of other types of projects that address the selection criteria and that have the strong prospect of producing 'cleaner, cheaper, smarter' results compared to the current system" (EPA, 1997b).

EPA's Pilot Project Examples

In industry proposals of facility-based XL projects, the EPA suggested that "[n]ational environmental requirements may not always be the best solution to environmental problems. Substantial cost savings can sometimes be realized, and environmental quality enhanced, through more flexible approaches involving pollution prevention." Accordingly, the EPA proposed that pilot projects could focus on individual facilities and test alternatives to current environmental management approaches driven by compliance with existing regulations. The

EPA sought proposals with overall objectives to devise and test more flexible approaches resulting in both better environmental results and reduced compliance costs (EPA, 1997b).

The EPA suggested that some XL projects might focus on national environmental regulations that apply to many industries. The EPA acknowledged that these regulations "are often promulgated piecemeal over a long period of time rather than as a comprehensive environmental program." Thus, one type of project to address this particular problem "might take the form of combining all federal (and possibly state) requirements for an industry into a single, integrated Final Project Agreement. Sector-based and place-based strategies might be combined in a project that focuses on a number of facilities in the same or related industries within a given geographic region or ecosystem." The EPA further suggested that "[p]rojects might propose development of enforceable 'best management practices' for pollution prevention or pilot the application of upcoming ISO 14000 voluntary environmental standards within a specific industry sector" (EPA, 1997b).

The EPA suggested that government agencies might sponsor projects that address the unique issues faced by governmental agencies. Those might include the optimization of environmental control strategies or the ability to reduce overall compliance costs by controlling specific pollution sources out of reach by government regulators.

Three companies' Project XL proposals have been used by EPA to illustrate successful projects. The following summaries were obtained from the EPA.

Intel Corporation

Intel's Fab 12 facility, which manufactures semiconductors in Chandler, AZ is implementing an *Environmental Management Master Plan* that includes a facility-wide cap on air emissions to replace individual permit limits for different air emissions sources. The Final Project Agreement was signed on November 19, 1996. In this agreement, Intel committed to:

- Maintaining air emissions for oxides of nitrogen, sulfur dioxide, carbon monoxide, particulate matter, and volatile organic compounds at a level that ensures the current facility, and any other manufacturing facility built at the site, is a "minor" air emissions source as defined by the *Clean Air Act*
- Using state health-based guidelines to establish enforceable emissions caps for emissions that affect the community adjacent to the facility; in addition, these health-based standards will be used voluntarily to set emissions levels to increase protection for those working in the facility
- Reducing water consumption and the generation of solid, nonhazardous chemical and hazardous waste
- Establishing property line setbacks twenty times greater than required by local zoning authorities
- Reducing vehicle miles traveled by employees
- Participating in equipment donation and training programs

Intel is the first company to agree to make all its environmental data publicly available on the Internet as part of a standard reporting mechanism.

INTEL CORPORATION

SUPERIOR ENVI-RONMENTAL PER-FORMANCE	The implementation of Intel's project in Chandler will protect the environment by: • Reducing up to 60 percent of the solid waste and up to 70 percent of the nonhazardous chemical wastes the facility generates by the year 2000 • Recycling up to 65 percent of the fresh water used at the facility • Balancing limits on hazardous air pollutant emissions with health-based guidelines
REGULATORY FLEXIBILITY	Regulatory flexibility will allow Intel to make operational changes without permit review, as long as permit limits are met, and the project includes multi-media, performance-based permits that specify performance levels for each regulated pollutant to be used at the new facility.
STAKEHOLDER INVOLVEMENT	Intel is working to ensure that stakeholders are involved in the environmental design and impact assessment of its proposal and are informed and have an opportunity to fully participate in project development. Efforts so far have included: • Establishment of a Community Advisory Panel to serve as a full partner in the project's development • A massive outreach effort to local citizens (including 25,000 hand-delivered notices) • The involvement of national, regional and local non-governmental organizations that provide substantial comments on the project • The use of Intel and EPA websites to increase the transparency of project development
APPROACHES TO BE TESTED	• The efficiency of performance-based caps in lieu of pre-construction review • The effectiveness of community involvement in decisionmaking as an incentive for improving environmental performance • The role of innovative technology (e.g., remote sensing and environmental monitoring) as an incentive for improving environmental performance • The value of incorporating nonregulated items into the regulatory permit process

Jack M. Berry Inc.

Jack M. Berry Inc., a mid-sized, juice-processing facility in LaBelle, FL is developing a facility-wide, comprehensive operating plan that consolidates 23 federal, state, and local environmental permits and all operating procedures into a single manual for the facility. The Final Project Agreement was signed on July 8, 1996.

The implementation of Berry's project will protect the environment by:

- Reinvesting cost savings into the voluntary installation of a new peel dryer used in fruit processing; this will result in a reduction of air emissions of volatile organic compounds, sulfur dioxide, and oxides of nitrogen
- Reduction of the number and types of solvents and lubricants used onsite and replacement with a number of environmentally friendly materials

Other voluntary goals will be incorporated into the final comprehensive operating permit that is still being developed, such as:

- Reduction of water use through reuse and more efficient management practices and technologies
- Reduction of solid waste generation through increased recycling

EPA offers flexibility to Jack M. Berry Inc. by providing the opportunity to use a comprehensive operating permit. This permit for the operation and regulation of the entire facility will maintain all environmental standards, and consolidate federal, state, and local facility permits. This eliminates the requirement of preparing multiple permit applications every few years — a benefit that will result in significant cost savings for the company.

JACK M. BERRY INC.

STAKEHOLER INVOLVEMENT	Jack M. Berry Inc. is working to ensure that those parties with a stake in the environmental concepts and impacts of its proposal are informed and have an opportunity to fully participate in project development.
	Efforts so far have included:
	• The development of a Stakeholder Committee, which includes representatives from the local chamber of commerce, a regional economic development group, local environmentalists, local representatives of two national groups (Audubon Society and the Nature Conservancy), the U.S. Department of Interior and the mayor of LaBelle
	• The review of the proposed project agreement by the Stakeholder Committee
	• A public meeting, conducted by the Stakeholder Committee, to inform and seek comment and input from all interested citizens about the development of the Final Project Agreement
	• Additional public outreach focusing on two local counties
	• The benefits and pitfalls of comprehensive operating permits that meld dozens of local, state, and federal permits into one
	• The impact of permit consolidation on costs and expenditures
	• The impact of permit certainty on cost of capital

Weyerheuser's Flint River Operation

Weyerheuser Company's pulp manufacturing facility in Oglethorpe, GA is striving to minimize the environmental impact of its manufacturing processes on the Flint River and surrounding environment by pursuing a long-term vision of a minimum (environmental) impact mill. Weyerheuser Company is taking immediate steps by decreasing water use and meeting or exceeding all regulatory targets. The Final Project Agreement was signed on January 17, 1997.

Through a combination of enforceable requirements and voluntary goals, Weyerheuser will improve the health of the nearby Flint River and surrounding watersheds by:

- Cutting its bleach plant effluent by 50% over a 10-year period
- Reducing water usage by about 1 million gallons a day
- Cutting its solid waste generation in half over a 10-year period
- Committing to reduce energy use
- Reducing constituents of hazardous waste
- Improving forest management practices in over 300,000 acres of land by stabilizing soil, creating streamside buffers, and safeguarding unique habitats
- Implementing ISO 14001 standards to create an effective environmental management system

EPA is offering Weyerheuser the flexibility to consolidate routine reports into two reports per year and to use alternative means to meet the requirements of new regulations that prescribe maximum achievable control technology. EPA also is waiving government review prior to certain physical modifications, provided emissions do not exceed stipulated levels.

WEYERHEUSER

STAKEHOLER INVOLVEMENT	Weyerheuser Company is working to ensure that stakeholders are involved in the environmental design and impact assessment of its proposal and have an opportunity to participate fully in the project's development. Efforts so far have included: • A series of regional public meetings in Oglethorpe, GA • Personal contacts through telephone calls and meetings • Oral briefings and broad distribution of written descriptions of Project XL to both management and staff employees • Oral briefings and the distribution of a written project summary to interested national, non-governmental organizations • An annual stakeholder public meeting to share Project XL performance data (scheduled for January 1998) • Publications of notices in courthouses and local newspapers to convey an open invitation and the date and time of the scheduled public meetings *continued*

APPROACHES TO BE TESTED	• How does a facility operate under an environmental management system with a minimum impact goal?
	• Can new technology to meet ambitious environmental goals be created by a company together with stakeholders and government agencies?
	• Can *closed loop* technologies achieve environmental benefits well beyond *end-of-pipe* approaches?

The EPA is encouraging other businesses to follow Intel, Jack M. Berry, Inc., and Weyerheuser by signing up for the program. According to one environmental manager of a large business, many businesses are interested in participating in the XL Program but are not ready to sign up. Here again, incentives need to be offered to businesses to dedicate human and financial resources to voluntarily reduce their environmental impacts and go beyond compliance with environmental laws.

A study of industry incentives for environmental improvement conducted by Terry Davis and Jan Mazurek of Resources for the Future on behalf of GEMI concluded that business incentives for involvement in the XL program are weak to begin with and the risks of litigation and other failures are high (GEMI, 1996). Presently, "there is frustration over the length of the project review process and confusion over the role of stakeholders; facilities receive conflicting signals from different levels of EPA staff; and environmental health and safety staff are having difficulty convincing corporate executives of the tangible benefits of the programs" (GEMI, 1996). The GEMI report concluded that while XL has fallen short of its original goal of providing firms with flexibility within the existing system, it still presents businesses with a unique opportunity to become actively involved in new innovative programs that change their business practices to reduce impacts (GEMI, 1996). A closely related experimental program from the EPA is intended to motivate companies to improve their manufacturing and industrial practices to minimize pollution and their use of natural resources.

THE COMMON SENSE INITIATIVE

On July 20, 1994 EPA Administrator Carol Browner announced the formation of a new program entitled the Common Sense Initiative (CSI). According to the EPA, this program is intended to bring together businesses, federal and state governments, environmental and environmental justice groups, and labor to take a fresh look at the way the environment is protected. The key themes of the program are establishing flexible and creative new methods to achieve environmental goals; whole-industry, whole facility approaches; greater multimedia focus; greater incentives for pollution prevention; stakeholder involvement to identify "cleaner, cheaper, smarter" solutions that are good for both the environment and the economy; and consensus building on how to support the environment and reduce compliance costs.

The EPA has 40 CSI projects underway in the following six industry sectors: automotive manufacturing, computers and electronics, iron and steel, metal finishing, petroleum refining, and printing. According to the EPA's statistics, these six industries comprise over 11% of the U.S. Gross Domestic Product, employ nearly 4 million people, and account for 12.4% of the toxic releases reported by all American industry in 1992.

The ongoing projects include:

- A group of metal finishers developing new procedures to reduce regulatory burdens on small finishers in return for superior performance
- A group of petroleum refiners rewriting rules to consolidate, streamline, and simplify air emissions reporting requirements
- Iron and steel manufacturers writing principles on how to clean up abandoned iron and steel brownfields and return them to productive uses
- The computer and electronics industry working on ways to eliminate RCRA barriers to pollution prevention, recycling, and water conservation
- Auto manufacturers creating more flexible regulatory approaches to reduce burdens on industry, a cleaner environment, and improved community participation in environmental decision making
- The printing industry developing a permit system for printers that provides operational flexibility, reduces pollution across all media (air, water, and land), and improves protection of workers, communities, and the environment

After two years of effort, an oversight group has adopted a series of environmental goals that need to be attained by the initiative. They include:

- Improving information and data collection
- Improving energy efficiency through projects with the U.S. Department of Energy
- Increasing recycling and eliminating barriers posed by RCRA
- Addressing water quality issues
- Improving community-based approaches
- Exploring the green track, involving an alternative regulatory system
- Discovering new methods to implement emerging ideas

In contrast to the ELP and Project XL, the CSI is experiencing greater success as a program because businesses are much more accustomed to searching for ways to reduce costs and burdens as opposed to defining in words and actions what is meant by the phrase "environmental leadership." The GEMI report on CSI concluded that despite some statutory and administrative difficulties, "CSI ostensibly represents an important first step in identifying better ways to manage industrial performance. The initiative has clearly helped to incubate some innovative ideas that may ultimately result in cleaner, cheaper and smarter environmental management strategies" (GEMI, 1996). As is shown in Chapter 17, some environmental regulations have had a beneficial effect on businesses that have been forced to discover new methods to comply with a complex regulatory scheme. If the Clinton Administration's experiment on common sense initiatives succeeds, the government should become less responsible for dictating outcomes and more flexible in designing environmental protection measures that improve the ecosystem without unduly burdening business.

CLIMATE WISE

Climate Wise combines a number of these theories into a comprehensive program to achieve industrial energy efficiency and pollution prevention. The program is jointly operated by the U.S. Department of Energy and the EPA. It was designed to assist the U.S. in honoring its international commitment to reducing greenhouse gas emissions to 1990 levels by the year 2000. Climate change prevention measures are the focus of international negotiations. In the U.S., 280 Climate Wise participants, representing 6% of industrial energy use, are actively involved in the program. Participants include large businesses, such as DuPont, AT&T, Georgia-Pacific, Fetzer Vineyards, Johnson & Johnson, and General Motors, and smaller businesses like DeBourgh Manufacturing in La Junta, CO. DeBourgh uses environmentally friendly products and technologies to manufacture state-of-the-art lockers for schools and other users throughout the U.S.

The intent of the program is for businesses to develop a partnership with the government to reduce waste and become more energy efficient. According to the director of the Climate Wise program, the government is not seeking to dictate any specific technologies or create any regulatory barriers to reducing operating costs. Rather, through programs such as Climate Wise, the government is providing businesses with free technological advice on altering production processes, switching to lower carbon-content fuels and renewable energy supplies, substituting raw materials, implementing employee mass transit, car pool or van pool programs, and auditing and tracking energy use for efficient improvements. It is estimated that by the year 2000, participants will collectively save more than $300 million and reduce more than 18 million metric tons of carbon dioxide equivalent.

The government has significant technical expertise from national laboratories such as the Nuclear Research and Energy Laboratory, Los Alamos, NM and the Allied Signal Kansas City Plant, Kansas City, MO in developing state-of-the-art, clean manufacturing processes; energy-efficient technologies; and alternative fuels. The government is deploying these experts on an "as-needed" basis to seven state pilot projects to work with businesses to improve energy efficiency and to reduce pollution.

Climate Wise began as a voluntary pledge program where businesses seeking to reduce operating costs and gain national recognition agreed to voluntarily reduce pollution. Participating businesses are now encouraged to publish a policy or statement on energy efficiency, establish an energy management team, set improvement targets, monitor and evaluate performance levels, and increase awareness of energy efficiency among employees. Businesses can voluntarily report their emissions reduction through the *Energy Policy Act's § 1605(b)* database and ensure eligibility for Climate Wise recognition. The following success stories should help business managers promote these ideas within their companies:

DuPont
DuPont has developed a Climate Wise plan to reduce HFC-23 and PFC emissions and improve overall energy efficiency in its manufacturing operations. According to the Climate Wise program, DuPont is already producing a savings of $31 million each year through low- or no-cost energy-efficiency opportunities.

AT&T
AT&T is forecasting that its Climate Wise commitments will save the business $50 million each year and offset emissions equivalent to reducing 170,000 tons of carbon dioxide by the year 2000.

Breyers Ice Cream
On a smaller scale, Breyers Ice Cream's Farmingham, MA plant completed a $3.5 million energy-efficiency project. The company cut energy costs by $425,000 per year through reduced electricity costs of 33%, saving enough electricity to light 1000 homes. The cost savings boosted the plant from its ranking as the business's worst, threatened with closure, to second best — boasting an expanded output and the addition of 40 jobs.

Greenfield Tap and Dye
In Greenfield, MA a factory instituted an energy-efficiency program which saved the plant from closure. Employees cut annual energy costs by $90,000 and saved $180,000 annually in avoided disposal costs. Production time was reduced from eleven days to one.

Very Fine Juices
Very Fine Juices in Westford, MA currently recycles 90-95% of its waste materials, which now saves the company $445,000 per year in disposal costs. Its newly designed cooling towers have reduced water usage by 50 million gallons annually.

Quad Graphics
Quad Graphics of Pewankee, WI reduced its waste ink output from 17 55-gallon drums per month to less than 1 drum per month at no cost through simple process changes. The payback was more than $500,000 per year in savings.

State Climate Wise Partnership Agreements

Twenty-one states have entered into Climate Wise partnership agreements to design practical, environmentally friendly relations with their local industry. These states and their local government allies are providing technical and financial assistance to 174 businesses in an effort to improve environmental performance and reduce abuse impacts.

THE MERIT PARTNERSHIP FOR POLLUTION PREVENTION

The Merit Partnership for Pollution Prevention (Merit) is a cooperative venture of the public and private sectors to develop and promote pollution prevention practices and technologies that protect the environment and contribute to economic growth (Reich, 1997). Merit brings together businesses engaged in manufacturing, banking, insurance, environmental risk assessment, accounting and auditing, and the government to facilitate demonstration projects. These projects will show how environmental management systems can be strengthened and implemented in different industries to achieve both improved environmental performance and economic competitiveness among businesses in the U.S.

Merit was created in 1993 when government and industry leaders were beginning to realize that it was both possible and imperative for government and industry to work together to achieve their respective goals of environmental protection and economic growth (Reich, 1997). Merit's first project involved the metal-finishing industry in California, where pollution prevention projects were implemented in seven small- to mid-sized metal finishing facilities. The companies used innovative technologies to reduce waste generation and recover materials from wastestreams for reuse and recycling. The results were published in technology transfer documents, videos, and technical workshops to promote pollution prevention to other metal finishers.

Merit is currently working on a project with a number of industry partners to implement ISO 14001 environmental management systems. One such partner, Northrup Grummon, already has an environmental management system in place and is now tailoring that system to comply with the ISO 14001 standard. Merit is reviewing this and other environmental management system pilot projects to determine:

- Are the environmental protections afforded by environmental management systems substantial enough to be promoted and considered by regulatory agencies in their interactions with industry?
- What elements of environmental management systems are necessary to assure compliance with environmental laws?
- Can an environmental management system improve environmental performance? If so, what elements of such a system are necessary to achieve that end?
- Are environmental management systems economically sound for small- to mid-sized businesses?

- Can the implementation of an environmental management system result in economic benefits to a company? If so, are those economic benefits the result of specific elements of the environmental management system?

- Can an environmental management system improve a company's compliance record? If so, should environmental management systems be used as an enforcement tool?

Two current projects are in the process of answering these questions (Reich, 1997). Merit is developing an environmental management system template based upon Northrup Grummon's ISO 14001 environmental management system. Once the template is completed, Merit will explore whether smaller facilities, some with compliance problems, can improve their compliance track records as a result of using the ISO 14001 template. Merit recognizes that smaller companies, which are suppliers and distributors for major manufacturers, are good candidates for demonstration projects. Those smaller companies often need assistance with regulatory compliance and encouragement in adapting to new requirements (Reich, 1997).

The other Merit project is exploring whether companies may be able to obtain financial benefits as a result of having an environmental management system. This project is focusing on the opportunity for businesses to get new types of insurance coverage, or reduced premiums on current coverages, as a result of having lowered their environmental risk through an environmental management system.

EFFECTIVENESS OF THE VOLUNTARY PROGRAMS

The most obvious question that arises after reviewing the EPA's voluntary programs is whether any of them have resulted in lasting environmental benefits and risk reduction. On March 19, 1997 the EPA's Office of Inspection General (OIG) submitted an audit report to the agency as to whether two older programs, the Radon and ENERGY STAR® voluntary programs, actually had achieved these two objectives. The OIG concluded that both programs provided an "impetus to overcome the barriers to energy efficiency and change consumer behavior. As a result, they were effective at achieving environmental benefits and reducing health risks. . ." (EPA, 1997b). The audit report also provides some insights into how the EPA's other voluntary programs can benefit business and result in similar reductions of environmental risk.

The voluntary Radon Action Program was established by the EPA in 1985 to ensure that the required technical knowledge about this deadly, naturally occurring gas exists and is accessible to homeowners, contractors, and state and local officials. In 1992, the Office of Policy, Planning and Evaluation concluded that while the Radon Program had made some progress in increasing radon awareness and testing, public information alone would not be sufficient to achieve significant, long-term risk reduction. Hence, the EPA recommended radon testing in real estate transactions and building radon resistant homes as cost-effective approaches in high risk areas (EPA, 1997b).

In 1993 the EPA initiated the ENERGY STAR® program as part of the Climate Change Action Plan to reduce greenhouse gas emissions through voluntary partnerships with businesses and public institutions. The ENERGY STAR® program was designed to get consumers and businesses to use more energy efficient products. The program had three goals:

- Increase market penetration of existing energy efficient products
- Ensure that manufacturers' and homeowners' investments in energy efficiency are cost effective and product quality is sustained or improved
- Change consumer behavior

ENERGY STAR® programs focused on encouraging energy-efficient technologies in specific areas such as lighting, office equipment, commercial buildings, and homes.

The OIG audit was limited to the Radon Program and three ENERGY STAR® programs: Office Equipment, Buildings, and Homes. The environmental benefits of the Radon Program showed that in high radon areas, radon awareness was at 78% and testing was at 13%. The ENERGY STAR® program for Office Equipment achieved, by the end of 1995, a savings of 2.3 billion kilowatt hours of electricity at 1300 pounds of carbon emission. The OIG stated that "the role of voluntary programs is to encourage the manufacture and consumers' acceptance of risk reduction in the market phase. By changing consumer behavior and effecting a market transformation, voluntary programs achieve lasting environmental results and reduced risks. As the market is transformed by new environmental innovations, the EPA will be able to reduce its program support and allocate its resources to other products or programs" (EPA, 1997a).

The OIG audit concluded that voluntary, or nonregulatory, programs can be an effective tool for reducing risk and achieving environmental results, as long as they demonstrate several good management practices. Those practices include:

- Planning
- Educating people about incentives
- Providing quality support
- Working with outside organizations
- Evaluating progress and making adjustments

These good management practices enabled the program to achieve environmental benefits of energy savings, pollution reduction, and reduction of radon exposure (EPA, 1997a). The audit results will undoubtedly encourage the EPA to continue its efforts to have businesses regulate themselves instead of using old-fashioned command-and-control methods to protect the environment.

Changing human behavior is perhaps the most critically important component of a successful voluntary program. The EPA believes that educating people about incentives is an effective way to get people to act, especially when information about the problem is not enough to get the desired result.(EPA, 1997a). "Financial and market incentives are strong motivators for consumers and corporations. The more value corporations and consumers place on the incentives, the higher the rate at which they will take the desired action, thereby decreasing risk" (EPA, 1997a).

In order to be successful, voluntary programs need commitments from outside organizations to get businesses and consumers to take a desired action. The EPA has never had nor ever will have sufficient resources and expertise to effect behavioral changes. It must rely on businesses and other nongovernment organizations that are interested in achieving the same environmental goals. Outside organizations became the key to communicating the Radon Program's message because the EPA's messages on radon had a limited effect on many audiences; the informational materials produced by bureaucracies were often untimely and generic, reducing the number of audiences they reached; the EPA had limited effective channels for sending out the information; and the outside organizations were more connected to the target audiences (EPA, 1997a). The OIG's audit validates the EPA's approach of using voluntary environmental programs to create greater awareness of environmental impacts and more opportunities to discover ways to protect the environment.

The EPA is now relying on outside organizations to take effective steps to reduce the risk of environmental noncompliance. By placing the burden on business to act the EPA is making a wise choice due to its limited resources and inability to communicate a positive message to every prospective business partner about protecting the environment. As we show in the next chapter, the EPA's record of prosecutions of environmental crimes is another method of changing human behavior about environmental responsibility. Environmental prosecution and stiff sentences for convicted felons do serve as deterrents. The more effective approach, however, is to convince businesses to create environmental management systems to provide long-term solutions to reducing the risk of noncompliance.

Chapter 15

AVOIDING THE RISK OF COMMIT-
TING AN ENVIRONMENTAL CRIME

In the first decade of federal criminal environmental enforcement (1982-1992), 232 companies of all sizes were convicted. Leading the list of companies that were charged were Exxon, Texaco, Nabisco, Ralston-Purina, Keebler, W.R. Grace & Co., Ashland Oil, Orkin Exterminating Co., Ocean Spray Cranberries, and Pennwalt (Thornburgh, 1991). These prosecutions sent a message that was heard by the corporate community: environmental crime was bad for business.

At the beginning of the 1990s, companies were being prosecuted at a ratio of 4 to 1 over individuals. By 1997 this ratio has reversed. Fortune 500 companies are nowhere near as vulnerable to prosecution for environmental crimes as they were in the late 1980s. There are several reasons for this abrupt change. Major companies were the first to institute early warning systems to prevent and detect violations of environmental law. Legions of articles were written about avoiding environmental crime which caught the attention of corporate managers with titles such as: "Doing Time for Environmental Crime;" "Behind Bars: Prosecutors Sting Corporate Executives;" and "Environmental Crimes and the Sentencing Guidelines: The Time Has Come ... But It is Hard Time." In 1991 Attorney General Richard Thornburgh of the Bush Administration described polluters as follows at an address to an environmental law enforcement conference:

> We are dealing with offenders who do some of the dirtiest work ever done to human health and the quality of life. They illicitly trade in sludge, refuse, waste and other pollutants, and they pursue their noxious concealments only for the sake of gain. Everywhere — on our land, in our water, even in the air we breathe — they leave their touch of filth. (Thornburgh, 1991)

Thornburgh's strong words had the support of the public. In one national poll in the 1980s, the public ranked environmental offenses just below murder, but above heroin smuggling, skyjacking, and armed robbery (Bureau of Justice Statistics, 1984).

Large companies listened to their environmental managers who heard the beat of the government's environmental war drums. These same companies took advantage of the government's enforcement policies that favor audits and disclosures of noncompliance in return for mitigation of civil fines and penalties, but no criminal charges. Early alarms regarding prosecutors running amuck using broad environmental statutes to ensnare the innocent turned out to be just that — alarms that woke up executives who had no desire to turn in their pinstripes for prison stripes.

Environmental criminal law remains a tricky area where the smallest mistake can lead to adverse consequences, and occasionally companies get hit with a megafine. A corporate environmental management system must incorporate a mechanism to reduce the risk of violations and mitigate overall civil and criminal liabilities. To understand how to create a risk-reduction plan, it is necessary to know what constitutes an environmental crime. The following is a summary of the criminal provisions of the four federal environmental statutes used most frequently by the government to prosecute environmental crime, to explain what the government thinks is an environmental crime.

THE RESOURCE CONSERVATION AND RECOVERY ACT (RCRA)

Violators of the provisions of RCRA are subject to criminal penalties if the acts are committed knowingly. Any person who knowingly transports, causes to be transported, treats, stores, disposes or recycles hazardous waste without a permit or in knowing violation of a permit is subject to a maximum term of imprisonment of 5 years and a maximum fine of $50,000 per day of the violation. A person who knowingly omits or gives false information in a compliance report, or who knowingly destroys, alters, or fails to file compliance documents, can be jailed for 2 years and fined up to $50,000 per day of the violation. The same penalties are imposed on any person who knowingly transports hazardous waste or used oil without a manifest, or who knowingly stores, treats, transports or disposes of used oil or listed hazardous wastes in knowing violation of a permit requirement or applicable regulation. The most severe penalty under RCRA is the knowing endangerment provision, which provides for a maximum prison term of 15 years and a fine of up to $250,000 for an individual who violates any of the sections of RCRA and knows at the time that he or she thereby places another person in imminent danger of death or serious bodily injury.

THE CLEAN WATER ACT (CWA)

The *Federal Water Pollution Control Act*, commonly known as the *Clean Water Act*, provides for the imprisonment of up to 1 year, or a fine of up to $25,000 per day of the violation, or both, for any person who negligently violates any of various sections of the statute dealing with discharges or disposal of oil and hazardous substances, or negligently violates permit requirements. If the violator acts knowingly, a sentence of up to 3 years may be imposed, and fines can reach $50,000 per day of the violation. Furthermore, the violator who acts knowingly, and at the time of the act is aware that he or she thereby places another person in imminent danger of death or serious bodily injury, faces a maximum of 15 years in prison and a maximum fine of $250,000. Businesses can be fined up to $1,000,000 under this knowing endangerment provision. Second convictions double the maximum punishment available for all violations.

THE CLEAN AIR ACT (CAA)

Under the *Clean Air Act* and its *1990 Amendments*, a person who knowingly violates any of several sections of the Act may be imprisoned for up to five years, or fined, or both. Knowingly making false material statements, failure to notify or report as required, or tampering with or failing to install monitoring devices as required is punishable by a maximum of two years in prison, and a fine. Knowing failure to pay a fee owed the U.S. under the Act is punishable by up to one year in prison and a fine. Similar to RCRA and the CWA, the CAA also contains a "knowing endangerment" provision, which punishes violators for known releases into the ambient air of any listed hazardous air pollutant, which the violator knew at the time thereby placed another person in danger of death or serious bodily injury. Knowing endangerment is punishable by a maximum 15 year prison term and a fine. If such a release is perpetrated negligently, however, the maximum sentence is one-year imprisonment and a fine.

CERCLA

The *Comprehensive Environmental Response Compensation and Liability Act* (CERCLA) imposes criminal penalties on persons who fail to report a release of hazardous substances, store hazardous substances, without notifying the EPA, destroy records concerning hazardous substances, or file false information in a sworn claim against the Superfund. A person who is convicted of violating these provisions may be sentenced to three years imprisonment, or five years in the case of a second conviction, and receive fines in accordance with Title 18 of the Federal Criminal Code. Under Title 18, the government can prosecute persons who make false statements to the EPA and government investigators in violation of 18 U.S.C. § 371 (conspiracy) and 18 U.S.C. § 1001 (false statements). The penalties for these offenses are 5 years imprisonment and up to $10,000 in fines.

Unlike bank robbery, skyjacking, and murder, environmental crime sometimes means entirely different things to different people. Compare these two descriptions of the same case that was prosecuted by the state of Ohio in the early 1990s. In *Ohio v. Stirnkorb,* the state convicted the operations manager of a hazardous waste landfill for violating Ohio's hazardous waste laws by unlawfully and recklessly failing to evaluate excess rainwater on top of a waste cell before ordering it pumped into an adjacent creek, an unauthorized location. The first description comes from a former environmental prosecutor, now a defender, and the second from the environmental prosecutor who handled the case.

The former environmental prosecutor said: the manager was acting during an emergency caused by a downpour, took measures to pump only clear water into the creek and colored water into holding ponds, acted under obscure and conflicting regulations, and no pollution of the creek was ever shown (Gaynor, 1993).

The environmental prosecutor said: The case of *Ohio v. Stirnkorb* shows how seriously the state of Ohio takes its environmental protection responsibility. This was not an accidental violation caused by an imprudent reaction to an emergency. The hazardous waste facility managed by the defendant had a stormwater management plan approved by the Ohio EPA. The plan required that stormwater be pumped to a holding pond, where it could be tested to determine whether it was contaminated. The defendant admitted he knew of the plan and had always used it previously. Stirnkorb also admitted he made a conscious decision not to use the method at the time for which he was cited. The only reason he gave for this choice to disregard the plan was that it was quicker to just pump the waste into a drainage ditch that emptied into a nearby stream (Muchnicki, 1993).

The *Stirnkorb* case and several other early cases caused debate as to whether prosecutors had too much discretion in charging environmental offenses. In the past several years, however, environmental cases are becoming more routine as both prosecutors and defenders became more accustomed to handling these types of cases. With more cases being filed, there are more decisions handed down by courts interpreting statutes. In the final analysis, what constitutes an environmental crime has become more predictable.

What are prosecutors looking for when they exercise discretion in charging an environmental crime? Keep in mind that criminal prosecutorial decisions generally are made on a case-by-case basis, and are frequently controlled by specific factual, evidentiary, and legal considerations (Levin, 1991). Prosecutors generally will want to review the nature of the wrongful conduct, the impact on the general public, the motive of the potential defendant, the criminal and/or regulatory history of that individual or company, and the effectiveness of the criminal option (Levin, 1991).

Four factors are considered in determining the effectiveness of the criminal enforcement option: deterrance, public protection, retribution, and remediation. In addition, one seasoned environmental prosecutor, Kent Robinson of Portland, ME has said that each case has to have a major impact on the environment. Criminal cases are not prompted by technical violations of those trying to implement complex regulations. Most cases involve deceit of some kind, such as lying to regulators. "There's no quicker way to get a criminal violation than to lie to regulators. It shows criminal intent."

ENFORCEMENT OF ENVIRONMENTAL LAWS

In July 1996 the EPA released its annual enforcement report entitled *Enforcement and Compliance Assurance Accomplishments Report FY 1995*. This report contains a summary of each of the criminal environmental cases brought by the federal government or resolved by a plea of guilty or a conviction following a trial for the most recent fiscal year. This report is a good guide in determining what the government considers to be wrongful conduct that should be prosecuted criminally. The report also provides information on how often environmental crimes are prosecuted in the federal system. It does not give summaries of state prosecutions which vary greatly in number from state to state. Some states like California that have had a substantial number of environmental prosecutions are more active than other states that have prosecuted next to none. The EPA has reported that 70% of all enforcement actions are undertaken by states, and state penalties for violations are "orders of magnitude" higher than those of the federal government.

The EPA levied $76.7 million in fines during 1996 according to an enforcement update released on February 25, 1997. Iroquois Pipeline Operating Company paid the lion's share of the fines, $22 million, for clean water violations in building a pipeline. According to the EPA, this figure was the second largest in history. The largest was the $1 billion fine levied against Exxon in the Exxon Valdez oil spill case. Steve Herman, EPA's Assistant Administrator for Enforcement and Compliance, stated that "strong enforcement underscores [the Clinton] Administration's commitment to protect our air, our land, our water, and our health."

To underscore this resolve, the Administration is seeking new criminal environmental laws that would allow prosecution of attempted environmental crime that is stopped before the pollution occurs. The legislation would also extend the statute of limitations for pollution control laws from 15 to 20 years. The Attorney General, in announcing the proposed legislation, said that enactment of the Administration's bill would "provide strong environmental protection, swift justice for polluters, and forge better enforcement partnerships among federal, state, local and tribal governments."

Only about 74 criminal cases were cited in the 1996 enforcement report, or about an average of 1.5 cases per state per year. The EPA does not list the cases that were either dismissed by judges or resolved by not-guilty verdicts by juries. Five of the cases summarized are simply updates of enforcement actions that were reported in earlier editions of the yearly enforcement report. The remaining 69 cases are a relatively low number, considering the amount of waste that is produced and disposed of each year in the U.S. The *Accomplishments Report* demonstrates that the federal government's environmental prosecutions, despite its rhetoric, are generally few and far between.

Most of the cases involve egregious actions by individuals who have decision-making authority, and hence, control, over the disposal of hazardous waste. Based upon the summaries, there is not much doubt that the convicted defendants were in fact guilty. The government highlights the most inculpatory facts and avoids mentioning the defenses in these case summaries, so it is difficult to tell if the defendants were under the mistaken belief that their actions were legal. The federal criminal environmental statutes require low levels of criminal intent necessary to support a criminal conviction. People who handle toxic wastes and pollutants are assumed by the courts of knowing the consequences of unpermitted discharges of such wastes into the nation's ecosystems. Prosecutions are invariably successful when people discard noxious wastes into the environment without a permit. The low standards of culpability have caused considerable angst among legal commentators and defense counsel, but they have been upheld repeatedly by the courts mainly because most environmental crimes involve outrageous facts.

For centuries a fundamental tenet of jurisprudence in the English-speaking world traditionally has been the requirement of establishing the knowledge of the defendant as a key element in a criminal proceeding. A noteworthy exception to this general rule has evolved judicially in cases involving the violation of statutes which were created in the public welfare. Although the public welfare doctrine originated in two cases dealing with the *Federal Food and Drug Act,* in the past decade courts have expanded the doctrine to include the various federal environmental statutes. In practice, this expansion has resulted in the reduction, or even elimination, of the requirement of actual proof of criminal intent in successful environmental prosecutions.

Several of the major environmental statutes (including RCRA, the CWA, and the CAA) explicitly include a knowledge requirement for a violation to occur. A question that has arisen frequently in litigation under these statutes, and in scholarly commentary, is whether the explicit statutory language used by Congress should supersede the common-law public welfare doctrine. Most courts that have addressed this issue have concluded that the congressional history of the statutes prevents the outright imposition of strict liability for most environmental violations. The same courts, however, permit a lessening of the direct knowledge or intent requirement and allow juries to draw inferences of knowledge from the facts established. Courts have greatly assisted the prosecution of environmental crimes by broadly construing the public welfare doctrine so that businesses with any impact on the environment are within its reach.

The second major weapon available to the government in environmental prosecutions is the responsible corporate officer doctrine, which is inextricably intertwined with public welfare legislation and a formidable tool to attach criminal liability to people that law enforcement agents, and others, euphemistically refer to as "higher-ups." The responsible corporate officer doctrine can effectively place criminal liability upon officers, directors, and any managers at any level of the company who know, or by virtue of their position have reason to know, that a violation has occurred or is likely to have occurred. The responsible corporate officer need not have directly caused, ordered, or even been involved in the conduct giving rise to the violation in order for personal criminal liability to be attached. That person need not, in fact, have any direct knowledge of the violation at all.

The scholarly debates over the reach of the environmental statutes has not subsided as the government has increased its efforts to prosecute environmental crime. A close look at the actual cases the government has prosecuted shows, however, that despite the debate, the actual criminal conduct has been very clear and the defendants remarkably guilty. Frequently, the disposal is so blatantly illegal and obviously deleterious to the ecosystem that defendants plead guilty simply to avoid an unnecessary trial and throw themselves, literally, at the mercy of the sentencing court. These cases generally involve deliberate disposal of hazardous wastes in rivers, oceans, wetlands, sewer systems, fields, and on abandoned properties. On occasion, there is a significant health hazard to people, in addition to severe damage to the ecosystem, created by the discharge of the waste. Penalties usually, though not always, increase with greater environmental impacts. Frequently, the disposal is accompanied by false statements made knowingly to the EPA and state environmental protection agencies by businesses or their environmental consultants. Environmental crime is occasionally committed in conjunction with other crimes, such as racketeering, mail fraud, or simple fraud. Also, environmental crime occurs when businesses adopt poorly conceived disposal plans.

1996 Environmental Enforcement Statistics

	Criminal Penalties Assessed	Civil Judicial Penalties Assessed	Administrative Penalties Assessed	Dollar Value of Injunctive Relief	Dollar Value of SEP's
RCRA	$ 8,076,000	9,066,018	7,771,104	60,988,395	14,173,682
CWA	62,214,200	19,790,390	3,441,475	576,958,752	5,219,412
CAA	5,202,800	30,885,091	2,439,633	205,580,107	16,973,165
CERCLA	2,000	2,369,491	708,057	452,148,874	267,675
TOTALS:	75,495,000	62,110,990	14,360,269	1,295,676,128	36,633,934

Source: Environmental Protection Agency, 1997

Highlights of Criminal Cases

Below is a brief summary of criminal cases in the categories just mentioned as the cases are reported by the EPA in the enforcement report. A discussion then follows as to whether an environmental management system would have prevented these types of crimes from occurring. In some cases, the government did not report the sentence or fine, either because the defendants were awaiting sentencing at the time the report was prepared, or for some other unstated reason. Sentences and fines are noted to the extent that the government mentions them in the report. Some of the cases are not final because those defendants are exercising their statutory rights of appeal of their convictions and sentences. The government has not mentioned in its report which defendants are appealing.

Disposal

- A plant manager and shop foreman received 27 months of imprisonment and the business was fined $1.5 million for illegal disposal of toluene into a trash dumpster following warnings by state officials. This disposal resulted in deaths of two nine-year-old boys.

- An individual pumped 5000 gallons of waste contaminated with gasoline into the city sewer system and was fined $20,000. This case was recently reversed and a new trial is expected.

- A business used an illegal bypass system to discharge pollutants into Cleveland's sewer system. The president was ordered to perform 200 hours of community service and provide training regarding proper waste treatment and disposal to all 55 of his employees.

- An individual, over a 15-year period, poured hazardous waste into a floor drain that discharged into an outdoor ditch. He was fined $750 and required to perform 50 hours of community service.

- A business discharged lead and copper into the local sewer system and was fined $40,000. The president was sentenced to five months of home confinement.

- A waste hauler and business were sentenced and fined for disposal of liquid restaurant waste into San Antonio sewers.

- An owner and operator of a wastewater treatment plant abandoned the plant, allowing the discharge of sewage into a residential area. He was sentenced to 30 months of imprisonment.

- A former superintendent of a city ordered two employees to bury nine electrical capacitors containing PCBs in a city landfill.

- A plant manager ordered an employee to bury 68 acid-containing drums.

- A retired program manager for a municipal health department was responsible for illegal storage of 100 barrels of different types of wastes.

- An individual was responsible for the removal of asbestos in the context of sale of commercial real estate.

- An employee disposed of formaldehyde in drums on another employee's property.

- A vessel discharged oily water into Miami River.

- A foundation was fined $8,000 for its role in asbestos removal at a historical building.

- The owner and president of a business was responsible for disposing of 40 drums of trichloroethane (TCE) near a river bank in rural Oregon.

- A business was fined $44,000 for disposal of contaminated wastewater into the Mississippi River without a permit.

- An individual was responsible for disposal of hazardous waste on a mountain in Vancouver, WA.

- An owner/operator was responsible for a precious metal recovery operation that discharged hazardous substances into a brook and wetlands. He was sentenced to 16 months and fined $650,000.

- A business was fined $80,000 for discharge of 80,000 gallons of used oil after a thief tried to steal a motor attached to a storage tank, rupturing the hose and resulting in a discharge of oil.

- A plant manager at a sewage treatment plant was responsible for discharge of raw sewage into the Kanawha River after neighbors complained of foul odors in a wetland area adjacent to the river.

False Statements

- An owner of a pesticide application firm was sentenced to five years in prison for falsifying data regarding the type of pesticide sprayed upon breakfast cereal grain, causing a $140 million loss to General Mills.

- Two managers of a wastewater treatment plant were responsible for falsification of discharge monitoring reports.

- A process engineer at a farm kitchen was responsible for filing false statements and concealing information relating to cooling water discharged into a creek.

- An employee of an environmental laboratory made false statements that analytical procedures had been performed on behalf a client.

- The president of a water filter manufacturing business was responsible for submitting false written statements in laboratory reports to the EPA.

- A business and assistant vice-president were responsible for making false statements and causing others to give false statements regarding asbestos contamination following a manhole explosion.

- A business and its president were responsible for false reporting, monitoring, and laboratory reports.

- A sanitation district and acting superintendent were responsible for false statements regarding operations of the waste treatment facility. The district was fined $35,000 and an employee was sentenced to 27 months of imprisonment.

- A business and two employees were charged by a federal grand jury with making false statements regarding emissions and production levels at a mill.

Environmental Crimes Committed in Connection with Other Offenses

- The president of an environmental firm was sentenced to 57 months of imprisonment for defrauding an insurance business, mail fraud, and obstruction of justice regarding the need for asbestos removal during the renovation of a hotel.

- Three individuals were responsible for violating asbestos NESHAP rules by using homeless men to strip asbestos and throw large quantities down an elevator shaft

- An employee of the U.S. Forest Service was responsible for disposal of hazardous waste containing lead and chloroform in toilets at a recreational site while participating in illegal manufacture of methamphetamine.

- A business and an employee were convicted of immigration and tax fraud and racketeering by bringing Filipino laborers to the U.S. and committing "a myriad of environmental crimes," including ocean dumping, and violations of the *Clean Water Act* and the *Rivers and Harbors Act*.

Ill-Conceived Disposal Plans

- A business, vice-president, and general manager were responsible for disposal of chromium-contaminated hazardous waste into a 40-foot diameter hole dug beneath its facility, creating a plume of contamination which threatened the water supply of their neighbors.

- A former highway administrator was responsible for the disposal of highway paint waste at a closed county landfill where drums were buried in trenches.

- A former owner of a business was responsible for abandonment of 80 drums of acetone and acetone still bottoms on rental property.

- A president of a corporation was responsible for disposal of chemical waste in a dumpster in a low-income minority neighborhood in the District of Columbia.

- A tugboat captain was responsible for losing a barge after a faulty repair on a cable, resulting in a discharge of 750,000 gallons of oil after the captain failed to notify the Coast Guard that the barge had broken loose and was adrift in an unknown area.

- A corporate secretary was responsible for disposal of hazardous waste where it was discovered that the plating business disposed of thousands of gallons of hazardous waste by abandoning its facility.

- A business and its president were responsible for disposal without a permit of chromium-contaminated drag-out waste directly into neighbors' drinking water supply.

- A former president of a textile business was responsible for disposal by abandonment of hazardous waste. He was fined $415,082 and sentenced to 6 months of home confinement.

- An employee was found guilty of disposing waste oil on an unlicensed barge.

- A former civilian supervisor at a military installation was responsible for removing asbestos-containing material from a building owned by his wife and burying it on government property.

- A scrap metal recycler was responsible for discharging 1000 gallons of oil into the Schuykill River in Pennsylvania while attempting to salvage a storage tank.

- A business was responsible for negligently violating the *Clean Water Act* after a landslide ruptured its pipeline, causing discharge of oil into the Allegheny River in western Pennsylvania.

- An owner of a tree trimming service disposed of refuse from this business in the bed of a creek, thereby avoiding landfill charges.

ACHIEVING COMPLIANCE THROUGH ENVIRONMENTAL MANAGEMENT SYSTEMS

Would an environmental management system have prevented any of these businesses or employees from violating environmental laws? To minimize environmental risks and exposures, businesses need to understand the consequences of their actions and their impacts on the environment. The government reports that one company was warned by state officials not to dump its toluene waste in its trash dumpster prior to the deaths of two small children who were asphyxiated by the chemical fumes. That company could have altered its disposal practices if it had known the consequences of its actions. Yet the business apparently was unwilling to act merely on the basis of a warning from state officials. It took the chance that its disposal methods would not cause a physical or environmental injury. This is deliberate risk-taking behavior that often results in negative and severe consequences to the company.

A systems approach to the minimization of risk could have provided the business with a range of possible consequences of the business's actions, including civil and criminal regulatory violations. Managers overseeing the disposal of waste could have prevented these accidental deaths by employing a properly implemented environmental management system. The system has to have the capability of internal controls to work effectively to eliminate noncompliance. Performance must also be independently reviewed to assure no breakdowns in the system. Risk-reduction environmental management systems have to be dynamic, as well, to change with new circumstances and problems faced by the business each day.

It appears from the government's summary that management actively participated in the illegal disposal in many of the cases cited in the report. Again, the defenses and mitigating statements made by the defense to the jury during the trial or to the judge at sentencing are not contained in the report. Nonetheless, an environmental management system is entirely useless if it will not be embraced by committed members of the management team. Businesses that have regulatory responsibilities need to have a carefully defined program that weeds out managers and employees who disregard the law, regardless of the consequences to business operations.

Law enforcement authorities rely on tips from neighbors in discovering environmental crimes. Public citizens frequently make complaints about what they perceive as environmental nuisances. Employees frequently report on managers and fellow employees who they think are violating the law. These employees may have been directed to participate in unlawful activities. The reports are generally made to local or state officials, OSHA, the EPA, or the FBI. Law enforcement authorities take these reports seriously and have been increasing the number of trained agents to investigate these types of tips. Since the *Pollution Prevention Act of 1990*, the numbers of federal environmental agents have been steadily increasing.

Employees generally notify authorities because they want "to do the right thing," to clear their own conscience if they were involved, or because they believe that their own inaction may be perceived as complicity with the illegal actions of the business. Employees also report on their businesses when they become disgruntled, often as a result of actions taken against them when they are fired, demoted or passed over for promotion. These employees become the eyes and ears of the regulatory authorities. An environmental management system can prevent violations of environmental law in the first place. Another important function, however, is for the system to serve as a reporting mechanism to responsible managers when incidents occur. If properly reported, documented, and explained, an employee may not perceive the event as an environmental crime. This is an effective way to avoid false or mistaken claims by employees against a business that is committed to detecting and correcting noncompliance.

Many of the EPA's criminal cases involve sole proprietorships, where the illegal action is taken by the business's sole employee. An environmental management system can be designed for the smallest businesses, including those that have only one employee. The program would amount to a checklist for an individual to follow in conducting daily business operations. From a small gasoline station operator or a family-owned dry cleaner to the largest multinational corporation, all businesses need some guidance in managing environmental risk and minimizing environmental impacts. Perhaps none of the individuals mentioned in the EPA's enforcement report would have altered their conduct on the basis of an environmental management system. However, given the alternative of spending months, if not years, in jail for senseless acts, even the most hardened and malignant midnight dumper perhaps would have some misgivings about perpetrating an obvious crime on the rest of society.

What about those businesses whose employees lie in reports to management and to regulatory authorities? Businesses and their environmental consulting firms have been prosecuted, as have environmental laboratories, for falsifying data and other information. Extensive computer networks, management oversight, and regulatory review are necessary to discover the false statements in official and unofficial documentation. It is also necessary to find ways to prevent these people from violating the law.

Incipient behavior often causes problems, like impossible sales goals which result in salespeople bribing purchasing agents to achieve those goals. To avoid creating unnecessary risks, like guaranteed delivery times that inadvertently cause accidents, management systems need to create positive incentives that foster good corporate citizenship and remove barriers to compliance that cause problems and accidents to happen because businesses will be held responsible for failure to exercise supervision over work performed on their behalf.

WHAT OCCURS WHEN ACCIDENTS DO HAPPEN?

On July 1, 1991 the U.S. Department of Justice issued guidelines designed to encourage businesses to engage in self-auditing, self-policing, and voluntary disclosure of regulatory violations by employees. The guidelines are called *Factors In Decisions On Criminal Prosecutions For Environmental Violations In The Context Of Significant Voluntary Compliance Or Disclosure Efforts By The Violator (Disclosure Guidelines)*. The government acknowledged that the significant increase in environmental prosecutions should not create a disincentive to or undermine the goal of encouraging critical self-auditing, self-policing, and voluntary disclosure. Accordingly, if businesses adopt internal procedures to reduce the risk of noncompliance and voluntarily disclose regulatory noncompliance before the government commences an investigation, it is at least theoretically possible that the prosecutors may be convinced to refer the case for civil rather than criminal action.

The *Disclosure Guidelines* must be read in conjunction with the EPA's audit policy (see Chapter 11) and with internal EPA memoranda addressing how the agency conducts criminal investigations (Devaney, 1994). The guidelines provide civil penalty mitigation for regulated entities that voluntarily discover violations, promptly correct violations, disclose violations, and cooperate in good faith with the EPA. Under the *Disclosure Guidelines*, violations must not be the result of criminal conduct by the regulated entity or any of its employees; discovery must be through a voluntary environmental audit or as a result of an environmental management system; the disclosure must occur before a government inspection or investigation; prompt correction must occur within 60 days or, if more time is needed, as expeditiously as practicable; any condition that may create imminent and substantial harm must be immediately remedied; and, measures must be undertaken to prevent recurrence.

In view of the serious penalties that have been imposed on businesses, their management and employees, those who are responsible for environmental regulatory compliance should pay strict attention to the *Disclosure Guidelines* as they give the U.S. Department of Justice broad discretion in deciding whether to institute a criminal environmental case. The U.S. Department of Justice designed this program to ensure that prosecutorial discretion is exercised consistently nationwide and to give the regulated community a sense of how the government will exercise its prosecutorial discretion. The *Disclosure Guidelines* set forth the factors that the Department will consider in determining whether to bring a criminal prosecution for a violation of an environmental statute or whether to proceed civilly.

The U.S. Department of Justice's *Disclosure Guidelines* for environmental violations bear close resemblance to the measures initiated by the government to prevent antitrust violations, procurement fraud, money laundering and other financial crimes, as well as other white collar criminal offenses. Regardless of the subject matter of the violation, the Department is urging the regulated community to engage in self-auditing, self-policing, and voluntary disclosure of employee violations of the Federal Criminal Code. Such measures are viewed as mitigating factors in the Department's exercise of criminal enforcement discretion. Businesses that make voluntary disclosures, cooperate with the government, and conduct environmental audits to ensure compliance with environmental laws and regulations will be able to make a convincing showing that criminal prosecution of the company is not warranted.

The first factor the government will consider in determining whether to prosecute the company is whether it made a voluntary, timely, and complete disclosure of the matter under investigation. Second, the *Disclosure Guidelines* provide that the attorney for the Department will consider the degree and timeliness of cooperation. Full and prompt cooperation is essential, whether in the context of a voluntary disclosure or after the government has independently learned of a violation. Third, under the heading "Preventive Measures and Compliance Programs," the *Disclosure Guidelines* require the Department attorney to consider the existence and scope of any regularized, intensive and comprehensive environmental management system; such a program may include an environmental compliance or management audit.

Fourth, the U.S. Department of Justice will determine whether there is "[p]ervasive noncompliance [which] may indicate systemic or repeated participation in or condonation of criminal behavior." The *Disclosure Guidelines* provide that "the attorney for the Department will consider a lack of a meaningful compliance program, ... the number and level of employees participating in the unlawful activities and the obviousness, seriousness, duration, history and frequency of noncompliance." Fifth, the *Disclosure Guidelines* state that "[e]ffective internal disciplinary action is crucial to any compliance program." Sixth, the guidelines require the attorney for the Department to consider the extent of any efforts to remedy any ongoing noncompliance. Significantly, "[t]he promptness and completeness of any action taken to remove the source of the noncompliance and to lessen the environmental harm resulting from the noncompliance should be considered."

Businesses that have strong environmental management systems and that make a timely, complete, and voluntary disclosure have the best opportunity to secure a declination. No business should consider such a course of action without an independent assessment by counsel. Time cannot be wasted in this process because the *Disclosure Guidelines* require a prompt disclosure that is truly voluntary; that is, not specifically required by law, regulation, or permit.

It has been suggested that before disclosing any violation, counsel should negotiate with the government and obtain either immunity from prosecution or a plea agreement to a lesser offense (Weisenbeck and Casavechia, 1992). Immunity for the business is unlikely because the government uses that procedure to further its investigations, not to terminate them. Pleading to a lesser offense defeats the whole purpose of voluntary disclosure, which is intended to result in businesses avoiding prosecution.

Before the decision to disclose can be made, it is necessary to determine what will be disclosed and the likelihood that disclosure will result in an indictment of the business or its employees. Counsel may seek to protect the results of audits by asserting state privilege, attorney-client privilege, the work product doctrine, or the critical self-analysis privilege (Weisenbeck and Casavechia, 1992). The *Disclosure Guidelines* do not require production of these reports, but businesses that express a willingness to make them available together with the names of all potential witnesses, will be treated much differently than those that do not. EPA surveys have found that approximately 50% of businesses that confront environmental risks are doing some form of environmental auditing. The most effective method of avoiding prosecution will be an independent, comprehensive environmental health and safety auditing program, coupled with voluntary disclosure.

The voluntary disclosure program is designed to place the maximum pressure on businesses to disclose the illegal activities of their employees in order to escape corporate criminal liability. At best, employees will face certain disciplinary action, including suspension or discharge for their activities. At worst, corporate disclosure of confidential information will result in ruined careers and criminal prosecution. Businesses that expose employees to government investigation and prosecution can be expected to face serious morale problems unless a system is employed to educate and train employees regarding their environmental responsibilities.

Notwithstanding these disadvantages, businesses should start to implement the criteria to shield themselves from prosecution. It has been suggested that businesses should institute educational and training programs at all levels to impress upon employees the importance of disclosur. (Weisenbeck and Casavechia, 1992). In addition, businesses should implement mechanisms for speedy analysis and reporting of any suspicious, environmentally related incidents and designate an office to be in charge of receiving all such reports. When problems occur, businesses should take all reasonable precautions not to overdisclose unnecessary problems for the business and its employees and make certain that early disclosures are not based upon incomplete or improperly analyzed information. Early disclosure based upon inaccurate information may not only bind the business, but may also limit the employees' options and put them at greater risk for either prosecution or disciplinary action (Ogren, 1991).

Even under the best of circumstances, when a disclosure of an environmental violation is made to the government, it will occur in a highly contentious and adversarial context. When enforcement activity is central to policy implementation by government agencies, the resulting relationship between the regulators and the regulated is one of mutual suspicion, distrust, and in some cases, open hostility (Pitt and Groskaufmanis, 1990). Any subsequent monitoring activity by the government or further investigation will add to the disruption of the business's daily operation. This tension has to be eliminated and mutually respectful relationships restored in order to minimize further disruptions. The more information employees have concerning the perils of environmental violations, the more likely this information will act as a deterrent to future illegal conduct.

U.S. SENTENCING GUIDELINES FOR
ENVIRONMENTAL OFFENSES

The current status of sentencing guidelines for environmental offenses is in flux. On November 16, 1993 the Advisory Working Group on Environmental Sanctions to the U.S. Sentencing Commission issued a working draft of recommended sentencing guidelines setting forth the criminal penalties for organizations convicted of federal environmental crimes. A sharp dissent from two members of the Working Group on the extreme sanctions for environmental crimes devised by the group has delayed the final promulgation of the guidelines (Guerci and Hemphill, 1993). The U.S. Sentencing Commission is reviewing the draft, and it is expected that some form of the draft will be submitted in 1998 to Congress for final approval. Like the U.S. Department of Justice, the Working Group has placed a major emphasis on companies' commitments to environmental compliance, cooperation, and self-reporting as mitigating factors in sentencing. According to the draft sentencing guidelines, if the court is convinced that such efforts have been made, the fine may be reduced.

The problem with the draft is that the aggravating factors outweigh the mitigating factors and allow for extreme fines and sanctions to be imposed on businesses that take extra, but unsuccessful, efforts to avoid noncompliance. Under the draft guidelines, if the court concludes that the company has benefitted by avoiding environmental compliance or avoided environmental compliance issues by failing to have a compliance program or other organized effort to achieve and maintain compliance with environmental requirements, and has not cooperated with the government or otherwise engaged in self-reporting, the court may use this information to impose an even stiffer sentence. The court may then increase the potential fine and punishment to the company, its officers, its directors, and its senior management personnel under any of these circumstances where aggravating factors are present.

Until the Sentencing Commission accepts or further revises the draft guidelines for environmental offenses, judges must use *Chapter 8 — Sentencing of Organizations* as guidance for sentencing organizations convicted of environmental offenses. That chapter provides that a criminal penalty can be reduced if a company has an effective program to prevent and detect violation of law.

Avoiding environmental noncompliance should be among the highest priorities of a company no matter how large or small. Proactive steps are important to reduce the risk as much as possible. The development of a document control system, as discussed in the next chapter, is a key component of an effective strategy to prevent environmental crime and to demonstrate environmental compliance, when necessary, to regulators.

Chapter 16

USING DOCUMENT CONTROL SYSTEMS TO REDUCE RISK

There is no better way to shut down a government investigation of an alleged environmental violation than to show a fully documented and verified record of compliance history. A document control system and the management of records are essential components of business operations. Sloppy control procedures can result in much aggravation and expense to the business when documents are not easily retrieved when needed. But daily operational problems pale by comparison to the charge of obstruction of justice when a prosecutor accuses a business of deliberate destruction or concealment of evidence. Document control is an important part of an environmental management system that can result in a significant reduction of environmental risk if set up and handled properly.

Problems are not limited to the rare criminal cases brought against businesses. In one civil case, Texaco discovered that document destruction by rogue employees was a contributing factor to immense civil liability and an onslaught of bad publicity. One employee tape-recorded others making disparaging remarks about minority employees who had sued the company. According to newspaper accounts, the recorded conversations also involved a discussion of destroying possible evidence. The company paid millions of dollars in settlement costs and had its reputation for equal employment severely damaged. A document control system that forms an integral part of a system's operations can address the issues which created Texaco's liability, achieve regulatory compliance, and thus avoid civil and criminal liability. ISO 14001 provides useful guidance on how to operate an effective document control system.

ISO 14001 GUIDELINES

ISO 14001 has two sections devoted to documentation: (1) environmental management system documentation and (2) document control. These requirements and ISO 14004 guidelines are discussed briefly in Chapters 4 and 19. The ISO 14001 documentation requirements now need to be considered in the context of adopting general methods of document control to reduce risk exposures.

Environmental Management System Documentation
ISO 14001 states under the heading "Environmental management system

documentation" that businesses shall maintain information in paper or electronic form to describe the core elements of the environmental management system. The ISO 14001 guidance for the standard provides: "The level of detail of the documentation should be sufficient to describe the core elements of the environmental management system and their interaction and provide direction on where to obtain additional information on the operation of specific parts of the environmental management system." The documentation can be in the form of a single manual, if desired, and may be integrated with documentation of other systems implemented by the organization. Related documentation can include process information, organizational charts, internal standards and operation procedures, and site emergency plans.

A written manual of the environmental management system can be readily disseminated to employees, suppliers, and other interested parties — and is strong evidence that a company is serious about its compliance efforts. The written policy should address the acceptable behavior of the businesses' employees and state the expectation that every employee will follow the policy.

Document Control

ISO 14001 provides a standard for businesses to establish and maintain procedures for controlling all environmental documents under the heading "Document Control." A document control system, however, includes procedures that require not just environmental documents but all relevant documents to be preserved for set retention periods until they can be destroyed consistently with laws and regulations and without damaging the interests of the business. First, we will discuss the ISO 14001 requirement to establish and maintain procedures for controlling all environmental documents and then how these procedures interrelate to a general records retention system.

The ISO 14001 *Standard for Document Control* provides a system for controlling document records to be established and maintained so that the documents can be located, periodically reviewed, revised as necessary, and approved for adequacy by authorized personnel. Current versions need to be made available at all locations where operations essential to the effective functioning of the environmental management system are performed. Obsolete documents should be assured against unintended use by labeling them as such or by promptly removing them. A decision may be made to suitably identify and retain certain records for legal and/or knowledge-preservation purposes.

The Document Control Standard also requires that documentation be:

- Legible
- Dated (including dates of revision)
- Readily identifiable
- Maintained in an orderly manner
- Retained for a specified period
- Traceable to the activity, product, or service involved

Procedures and responsibilities need to be established and maintained concerning the creation, modification, and disposition of the various types of environmental records. Training records and the results of audits and reviews need to be stored and maintained so that they are readily retrievable and protected against damage, deterioration, or loss. Retention times must be established and recorded.

The ISO 14001 requirements regarding records provide that the procedures for identification, maintenance, and disposition of records should focus on those records needed for the implementation and operation of the environmental management system, and for recording the extent to which planned objectives and targets have been met. Environmental records can include, but are not limited to, information on applicable environmental laws or other requirements; complaint records; training records; process information; product information; inspection, calibration, and maintenance records; pertinent contractor and supplier information; incident reports; information on emergency preparedness and response; information on significant environmental aspects; audit results; and management reviews.

When an effective environmental document control system using ISO 14001 is implemented, a business can comply with relevant laws, save substantial sums of money by eliminating valueless papers, and improve access to information for review. A well-crafted document control policy statement assists senior management in understanding the rationale for improving records management procedures. It also ensures that the system, as implemented, achieves broad corporate environmental compliance principles and goals.

A comprehensive environmental document control system cannot be created independently from a business's other document control procedures. Indeed, all sizes of companies already store vast quantities of records for business purposes wholly apart from efforts to achieve environmental management system conformance. The existing records procedures need to be systematically applied to the environmental document control program to ensure that the proper records are preserved for the correct period.

PRACTICAL STEPS TO AN ISO DOCUMENT CONTROL SYSTEM

Here are some practical steps a business can take to meet the requirements of ISO 14001 by incorporating an effective environmental document control system into the business's existing records system:

Inventory
A business should begin the process by undertaking a comprehensive records inventory. The inventory should involve all departments to determine what records the business creates on a daily basis in the ordinary course of business, and what records it maintains to ensure that the program covers all major documents.

The records inventory may serve a dual purpose. It should disclose which employees and agents are not exercising caution in their documentation of the business's internal and external business activities. It may also uncover documents most likely to cause concerns regarding potential liability. The inventory can be designed to locate all the records the business generates. It should take into consideration the structure of the organization and the scope of its activities (the states and foreign countries in which it does business), and the agencies that regulate its activities. The types of documents located in the inventory can be summarized in a series by content such as accounts receivables, contracts, invoices, personnel files, and quality control, etc. (Skupsky, 1994). Finally, all federal and state laws and regulations that apply to the business's activities must be researched to determine what records need to be retained and for how long.

Determination of Retention Periods

After conducting initial research, one expert recommends the identification and use of functional retention categories to establish a legal document retention schedule. He advises that there are five legal categories that affect records retention periods:

- Requirements to keep records, such as certain employment records, that do not state retention periods
- Requirements to keep records for specified retention periods, such as hazardous waste manifests
- Statutes of limitations or limitations of action periods that specify when legal actions or lawsuits can be initiated
- Limitation of assessment periods that specify when taxes may be assessed and tax records audited
- Pending, threatened, or potential litigation and government investigations (Skupsky, 1994)

The business must also determine how long to retain each category of documents.

The next step would then be to assign one of the functional records categories listed above from the records retention schedule to each record listed in the inventory. The records retention period for the functional retention category also becomes the record retention period for each of the document series (Skupsky, 1994).

Many records should be kept not only because of requirements of the law, but also because the organization may need the records to defend itself in regulatory or third-party litigation or for other compliance purposes. Many companies facing environmental liability, for example, have discovered that old insurance policies and secondary evidence of insurance coverage (such as invoices, cancelled checks, and correspondence) are valuable in obtaining indemnification for past and future clean-up costs. Other companies have conducted historical records reviews to discover that predecessor and affiliated organizations may be responsible for the disposal of waste byproducts. Records may be needed for a company to prove its own case as a plaintiff in a contract action and in many other types of civil litigation. A search should be made of contractual risk transfer and indemnity agreements which could limit or transfer risk.

There are other reasons to retain some records for an indefinite period of time. Each business should conduct a risk analysis to determine for itself the period of time records should be allowed to exist, as well as the location for storage. Losing valuable documents is inexcusable and destruction is irreversible.

Companies need to make careful and reasoned predictions regarding the likelihood that certain records may be useful in the future to establish a record of compliance, to defend against claims of civil or criminal liability, and to keep an accurate and complete record of business operations.

Development of Functional Records Retention Schedules

Once the legal basis for the retention of records is established, then a functional records retention schedule can be fully developed. There are several ways that this can be done. In most cases, businesses have to tailor the system to meet their particular needs. One possible approach is to divide the schedule into five essential steps:

- Sort the document inventory into related groups
- Develop initial retention categories, codes, and descriptions based upon the organization's needs and the scope of its business
- Revise the retention categories until each entry from the document inventory can be assigned to the appropriate retention category
- Sort the retention categories in order of the retention codes
- Assign the appropriate legal group code from the legal group file (Skupsky, 1994)

Other methods can accomplish the same objective if they are carefully planned and orchestrated by employees who have common sense, good judgment, and a decent understanding of a company's compliance requirements and objectives. Whatever system is used, however, the file information must be readily accessible — and on short notice.

Confidential Documents

Businesses need to be sensitive to issues involving the accessibility of confidential material. A document control system should include procedures on how to handle the creation, retention, and dissemination of confidential documents on disk or other forms of electronic communication. This will ensure that confidential documents do not get disseminated inadvertently and cause the business embarrassment or liability.

Integration into the Compliance System

A good document control system alone is insufficient to assure that a business will achieve regulatory compliance on all document control issues. Businesses must adopt procedures to integrate this document control system into an overall compliance system. This may begin with a one-page statement of business policy distributed to all employees regarding the reasons why accurate records, properly preserved, are an important component of business operations. For larger companies, a document control committee, made up of representatives from the legal and other departments of the business, can develop the policy and provide continuing advice regarding its interpretation (Kaplan, Murphy, and Swenson, 1994).

Employee Training

Businesses also need to train employees on how to discard unnecessary materials. Employees need to be restrained from the creation of unnecessary documents and be encouraged to establish the business's credibility by documenting the compliance program. Employees also need guidance on how to write documents that will avoid unnecessary legal consequences in the future. If records are written in a careless fashion with hyperbole, legal conclusions by non-lawyers, speculation and offensive language, a document control system will result in harmful evidence that can prejudice a judge or jury against the business. This is particularly true for e-mail, where employees tend to write informally and frequently do not understand the consequences of ill-chosen words and careless messages.

Audit

For any program to be effective, it should have a reliable audit component to identify employees who tend to save and train them regarding what should be retained and discarded. Written materials about the document control system should be distributed to each employee with specific instructions regarding the legal risk and increased costs of over-retention of documents. Over-retention may decrease computer storage capacity and add to storage and warehousing costs. Over-retention of records can add legal costs in the event of litigation without contributing to compliance efforts. An interdepartmental document control committee should have at least one member, and preferably more, who are well versed in computer technology to streamline records procedures.

THE EFFECT OF NEW TECHNOLOGIES

Because of employees' high usage of word processors, scanners, facsimile machines, electronic mail, and other technological advancements in recent years, companies must reassess their existing document control systems to determine whether new technologies are creating additional problems and how to solve these problems.

In a recent survey of large companies, an average worker received or sent 177 messages of all kinds (electronically, telephonically, or by mail) each day. The vast production of information and the time it takes to read and process information essential to business operations creates enormous problems for businesses. While computers have created many benefits for businesses, they have also caused a whole new set of problems which must be solved to make the business run smoothly. In the document control context, waste procedure improvements need to occur so that documents are not created and stored for future unintended users.

Many companies are purchasing new technologies that combine computers, facsimile machines, copiers, and printers to cut down on the unnecessary production of paper and the use of electronic energy. Service businesses like law firms, insurance companies, and financial institutions are using vast quantities of paper daily and need to address how to create new mechanisms to decrease their dependency on paper products and increase reliance on electronically stored information. When companies are equipped with CD-ROM technology, unnecessary space like libraries and file rooms can be eliminated when books and documents are put on disks. This can cut down overhead and simplify and speed up the retrieval of information. Within a number of years, paper may be an outdated form of recordkeeping.

Some companies are now solving the document-glut issue by storing information on document-imaging systems and destroying unimportant documents. Imaging systems provide an electronic alternative to the physical filing and storage of documents. A single, 12-inch, high-density optical disk can store approximately 100,000 pages of documents.

Until recently, these new technologies have not been universally accepted for records storage purposes. On July 20, 1994 the National Archives and Records Administration (NARA) authorized the limited use of the CD-ROM medium for the transfer of permanent records to the National Archives. The NARA recognized that a growing number of federal agencies are using optical disk systems for the management and dissemination of federal information. Businesses should seriously consider investing in such storage systems and refining document control procedures to take advantage of these new technologies. They also need to be reminded that today's state-of-the-art computer technology will be outdated in the 21st Century. It will be important to monitor the development of technological advances in document control systems to make certain information stored will be able to be retrieved in the future.

PERIODIC REVIEW OF THE SYSTEM

An effective document control system is dynamic, like an environmental management system, and needs to be reviewed and revised as circumstances develop. Businesses must also be careful when amending their document control systems once they are established. Any changes to the system occurring during litigation will most likely be perceived negatively by the judge or jury. Judges have frequently imposed sanctions on lawyers and their clients for document misdeeds in the course of civil discovery.

If businesses follow these methods and obtain additional advice on how the records control procedures can be effectively integrated into an environmental management compliance system, then the records that are retained will be more easily located. Those records will then be more valuable to the business when it is in litigation or needs immediate access to respond to problems.

Environmental management systems that implement document control mechanisms effectively can improve productivity. Problems can be identified and corrected before they escalate into regulatory violations. Environmental management systems can expose inefficiencies in production and service, as well as provide information that can be useful to management to make innovations to eliminate unnecessary waste and reduce sources of pollution. As discussed in the next chapter, companies are discovering new solutions to environmental problems based on a comprehensive assessment of environmental impacts, responsibilities and risks.

Chapter 17

ENVIRONMENTAL PERFORMANCE INNOVATIONS THAT REDUCE RISKS AND LIABILITIES

Through the implementation of environmental management systems, businesses can find new, effective ways to manufacture more environmentally friendly products and provide better services while reducing environmental impacts. Process and production innovations can diminish pollution, eliminate unnecessary wastestreams, and minimize energy consumption and usage of raw materials and natural resources. Performance innovations can also eliminate or reduce environmental risks, such as toxic releases into the environment. Manufacturing and service companies need to think about how to change their daily operations that impact the environment and improve delivery of green products and services. Aside from the obvious benefits to the environment, there are some good business reasons for these innovations. Environmental management systems can identify the obvious problems, but more importantly, can point the way towards discovering solutions to environmental problems, measuring success, and ensuring that the business remains profitable in the face of escalating environmental responsibilities.

In their book *The Total Quality Corporation*, Francis Mcinerney and Sean White studied ten major companies that have changed their operations to reduce pollution. They found that "zero waste equals zero defects. Those companies that pollute the least have the highest quality products and services." Waste or pollution, they further found, "is a key indicator of unnecessary cost, and, most important, inattentive routine management." Their book describes how these internationally successful companies improved their market share and lowered costs by redesigning systems and drawing connections between waste, quality, and margins.

This chapter will summarize some of these success stories and show how other companies are similarly gaining market share by finding innovative solutions to environmental problems. These solutions result in better product design, less waste, and lower operational costs. Many have occurred primarily as a result of our complex regulatory scheme. Environmental regulations can produce positive changes in how we do business. The change has to start somewhere, and in many of these companies it has occurred because the CEO and top managers became committed to responsible and profitable handling of environmental issues. Besides top management support, these companies have the following common elements in their approaches to environmentally based efficiency. They all:

- Have drawn the connection between waste and quality
- Use the waste-quality relationship to drive management decisions
- Use waste reduction to remove inefficiencies in customer service
- Use the environment as a catalyst to force organizational issues to the surface, where they can be solved
- Use environmental measures to increase shareholder value (Mcinerney and White, 1995)

All these companies have mature environmental management systems, fully integrated with systems operations, that can be used to identify situations in which innovative solutions are possible. Even law firms are beginning to realize that they too can be more proactive in improving performance by using total quality management practices. The following are some examples of companies that have responded positively to environmental regulations by discovering ways to reduce their environmental impacts using an environmental systems approach.

EXAMPLES OF COMPANY INNOVATIONS

Top Management Commitment

Ray Anderson, CEO of a large commercial carpet manufacturer, retained consultants who created the Evergreen Lease, the world's first perpetual lease for carpet. Anderson's company makes the carpet, installs it, maintains it, and replaces worn and damaged areas (one 18-inch square at a time), and recycles the used carpet tiles. The title for the carpet tiles never passes to the user; it stays with the manufacturer, along with the ultimate liability for the used-up, exhausted carpet tiles (Anderson, 1995).

Other companies have hired "eco-gurus" to give them sustainable-development makeovers. Ontario Hydro, one of the largest electric utility systems in North America, brought in two consultants at the request of its new CEO Maurice Strong, former secretary general of the RIO Earth Summit. The consultants produced a report recommending 98 changes and improvements in the operating systems; 40 have already been implemented (Frankel, 1995).

The Massachusetts Institute of Technology's Sloan School of Management has an Organizational Learning Center which positions learning as every organization's central activity and as a process which requires the active and continuous participation of all employees (Frankel, 1995). Anthony D. Cortese, Chief Executive Officer of Second Nature, a nonprofit organization, is working toward the vision of a world in which environmentally sound action will become "second nature" to all citizens (Cortese, 1994). Cortese is working with educational institutions throughout the U.S. to change how they teach to incorporate environmental education into curricula so that future scientists, engineers, and business people will design technology and economic activities that sustain, rather than degrade, the natural environment, enhance human health and well-being, and mimic and live within the limits of natural systems (Cortese, 1994). Cortese believes that "[a]ll professionals should understand their connection to the natural world and to other humans globally. They should know where products and services come from, where wastes go, and what they do to humans and other living species" (Cortese, 1994).

Pollution Prevention Studies

Cortese and others have urged businesses to conduct internal reviews to determine their global and local environmental impacts. Studies to determine how pollution can be prevented can have a profound impact on the company's, and industry's, operations. Michael Porter and Claas van der Linde have been collaborating since 1991 with the Management Institute for Environment and Business to explore the central role of innovation and the connection between environmental improvements and resource productivity. According to Porter and van der Linde, "the data clearly shows that the costs of addressing environmental regulations can be minimized, if not eliminated, through innovation that delivers other competitve benefits" (Porter and van der Linde, 1995a). For example, a recent study of 10 manufacturers of printer circuit boards produced 13 major changes: 12 resulted in cost reduction, 8 in quality improvements, and 6 in extension of product capabilities. Another study of activities to prevent waste generation at 29 chemical plants found that of a total of 181 waste prevention activities, only one resulted in a net cost increase. Many resulted in increased product yields (Porter and van der Linde, 1995a).

Ciba Geigy conducted a comprehensive study of its wastewater streams at its dye plant in Tom's River, NJ. The company made two changes to reduce pollution. As a result, the company increased its process yields by 40%, realizing an annual cost savings of $740,000. Similarly, an organic chemical company hired a consultant to study its reduction opportunities in its 40 wastestreams. The consultant's audit found 497 wastestreams (Porter and van der Linde, 1995b). Studies sponsored by the EPA in paper and pulp, paint and coatings, electronics manufacturing, refrigerators, dry cell batteries, and printing inks prove that innovation offsets to environmental regulation are common and that even in an atmosphere of intense regulation, the offsets can sometimes exceed the cost of compliance (Porter and van der Linde, 1995b).

Process Changes

The Dutch flower industry created a closed-loop reproduction process to avoid contamination of soil and groundwater with pesticides. The Dutch now grow their flowers in water and rock wool in advanced greenhouses, thereby reducing the need for fertilizers and pesticides. Water is recycled as it circulates through the closed-loop system (Porter and van der Linde, 1995a).

At a nylon plant in France, the Rhone-Poulence company invested 76 million francs to install new equipment to sell byproducts of the manufacturing process that were formerly considered wastes. The sale of the byproducts has

generated 20.1 million francs annually for the company. Dow Chemical redesigned its production process at one facility, which decreased its caustic waste
by 6000 tons a year and its hydrochloric acid waste by 80 tons per year. The
changed process resulted in an annual savings to Dow of $2.4 million for a
$250,000 investment (Porter and van der Linde, 1995a).

New Ideas for Old Products
    In 1990 Aeroquip Corporation, a $2.5 billion manufacturer of hose fittings
and couplings, decided that its products might be useful in meeting the need to
reduce waste and prevent pollution. "Aeroquip never thought of itself as a pro-
vider of environmental solutions," yet in the past several years it has generated a
$250 million business by focusing its attention on developing products that re-
duce emissions (Hart, 1997).

New Products
    Dupont's agricultural products business developed a new type of herbicide
that has led to a one billion pound reduction in chemical waste produced in the
manufacture of agricultural chemicals. Monsanto has developed a bioengineered
potato that defends itself against the Colorado potato beetle (Magretta, 1997).
Monsanto is predicting that widespread adoption of the product will eliminate
millions of pounds of pesticides and residues that are produced each year to
exterminate the beetle (Magretta, 1997). Monsanto is also genetically engineering cotton plants to produce Bt. bacteria, which are fatal to the destructive cotton
bud worm but harmless to humans and other living organisms. It is also producing Roundup® Herbicide, that, when sprayed on fields, kills weeds and eliminates the need for plowing — thus reducing soil erosion (Magretta, 1997). "The
Roundup® molecule has other smart features that contribute to sustainability. It
is degraded by soil microbes into natural products such as nitrogen, carbon dioxide, and water. It is nontoxic to animals because its mode of action is specific
to plants. Once sprayed, it sticks to soil particles; it doesn't move into groundwater, like a smart tool, it seeks out its work" (Magretta, 1997).

Co-location and Waste Exchanges
    BASF is designing and building chemical factories in China, India, Indone-
sia, and Malaysia that are less polluting than in the past. These facilities are
being co-located with others to create industrial ecosystems in which the waste
from one process becomes the raw material of another. Co-locations of these
industries is solving a common problem where recycling waste is not feasible
because transporting it from one site to another is dangerous and costly. In the
U.S., waste exchanges of hazardous materials are allowing companies to trade
their wastes with other companies that can use them in production processes by
advertising on the exchanges strategically located throughout the country.

Recycling
    In 1991 Japan promulgated new regulations to promote increased recycling

of products. Appliance manufacturers responded by redesigning products to reduce disassembly time (Porter and van der Linde, 1995a). Hitachi, the fifth largest company in the world, set three product recycling goals:

- Reduce disassembly time by half
- Recycle 30% more of its office products and household appliances
- Cut polystyrene packaging by half (Mcinerney and White, 1995)

Hitachi designers reduced the number of parts in a washing machine by 10% and the number of parts in a vacuum cleaner by nearly 40% (from 209 to 132).

In 1992 the German government required automobile manufacturers to take-back their cars for recycling. This encouraged BMW to create the first design for disassembly in the auto industry (Hart, 1997). Germany has a strict recycling law that requires manufacturers to take back from consumers all packaging that they use (Mcinerney and White, 1995). In the U.S., Amory Lovins of the Rocky Mountain Institute has designed a fully recyclable hypercar that is 30 times more energy efficient and 200 times cleaner than existing cars. The hypercar uses lightweight composite materials, fewer parts, virtual prototyping, regenerative braking, and very small hybrid engines (Hart, 1997). Hypercars have been described as being computers on wheels with microchips that "render obsolete most of the competencies associated with today's auto manufacturing, for example, metal stamping, tool and die making, and the internal combustion engine" (Hart, 1997).

The Dunlap Tire Corporation and AKZO Nobel developed a new radial tire that uses an aramed fiber belt instead of a conventional steel belt. "The new design makes recycling easier because it eliminates the expensive cryogenic crushing required to separate the steel belts from the tire's other materials" and improves gas mileage and safety as the new design improves the traction control of antilock braking systems (Hart, 1997).

Xerox Corporation has developed an asset-recycling management program that uses leased Xerox copiers as sources of high-quality, low-cost parts and components for new machines. Xerox has estimated that it saved between $300 million and $400 million in 1995 in raw materials, labor, and waste disposal by using refurbished parts for new machines (Hart, 1997). Conversion Technologies has created a unique process to recycle cathode ray tubes from computer terminals and television sets that were previously landfilled. The technology allows the CRT glass to be reused in electronic products or in colorful tile products.

Redesign to Avoid Regulatory Control and Eliminate Waste

3M improved its resource productivity when it was faced with new regulations to reduce solvent emissions by 90%. 3M reengineered its production processes to eliminate the solvents by coating the products with safer, water-based solutions. It also developed new techniques to run rapid quality tests on new batches of products that reduced hazardous wastes by 110 tons per year and yielded an annual savings in excess of $200,000. Raytheon similarly substituted a cleaning agent into a closed-loop system, which improved average product quality and lowered operating costs (Porter and van der Linde, 1995b).

Northern Telecom, a Canadian company that builds computerized telephone exchanges, eliminated CFCs from printed-circuit-board production in response to regulations that will ban CFCs worldwide by the year 2000. Northern Telecom's engineers experimented with a variety of other solvents to replace CFCs in the cleaning process when they decided to eliminate the cleaning process altogether. The company was able to tighten up production tolerances thereby eliminating the need for cleaning, saving more than $50 million in direct costs by the end of this decade (Mcinerney and White, 1995).

In 1991 a group of auto manufacturers agreed to help reduce emissions of over 65 toxic pollutants in the Great Lakes region. The program has successfully decreased emissions by 15%, and is being expanded to automobile manufacturers in other parts of the country.

Service Industry Innovations

The Russell, McVeagh, McKenzie, Bartleet & Co. law firm in Wellington and Auckland, New Zealand became the largest law firm to obtain ISO 9001 certification to total quality management standards. Russell, McVeagh's objective was to improve the effectiveness, flexibility, and competitiveness of the firm using ISO 9001. The law firm spent 18 months investigating the best practices and models to adopt and then measured their findings against key legal practice areas to establish standards that were able to be continuously monitored and upheld. An independent standards organization certified that the firm had implemented a quality assurance system subject to continuing audit for the design, development, production, and servicing for the provision of legal services for its offices in both locations.

Sullivan and Worcester, a law firm in Boston, has endorsed the *CERES Principles*. Other companies are taking active steps to reduce their environmental impacts. Suggested methods include car pooling, van pooling, and the development of alternative transportation plans that cut down on single occupancy vehicle trips to and from the office; implementation of the EPA's Green Lights program; reduction of paper usage and disposable paper food items, such as plates, utensils, and cups; improvement of food storage systems; integration of e-mail, facsimiles, copiers, and computers to improve performance; reduction of unnecessary data and paper collection; and promotion of recycling wherever possible.

PART THREE:

IMPLEMENTATION OF RISK MANAGEMENT AND RISK REDUCTION STRATEGIES

Chapter 18

GAPS ANALYSIS

Many large businesses have developed custom-designed environmental management systems, often without significant assistance or guidance from management consultants. The ISO 14000 series of standards can assist those businesses in further improving their environmental management systems or assist others in developing systems in accordance with internationally recognized standards. ISO 14001, titled *Environmental management systems — specification with guidance for use* (ISO 14001:1996 (E)), is the actual standard to which a business is certified. ISO 14004, titled *Environmental management systems — general guidelines on principles, systems and supporting techniques* (ISO 14004:1996(E)) is the guidance document supporting the ISO 14001 standard. These standards were released on September 1, 1996, and are available for purchase in the U.S. through the American National Standards Institute (ANSI) and the American Society for Testing Materials (ASTM). A properly designed and implemented ISO 14001 environmental management system can result in a significant reduction in environmental risk.

ISO 14001 has three distinct phases of implementation. Phase One is a gaps analysis, in which the business's existing environmental management programs are compared to the requirements of the ISO 14001 standard. This chapter presents one method of conducting a gaps analysis. Phase Two is the development of an environmental management system, following the requirements of ISO 14001, and then the implementation of the system. Phase Three involves the third-party assessment, or registration audit of the business, after the environmental management system is implemented. Chapter 20 discusses how the registration assessment is conducted.

WHAT IS A *GAPS ANALYSIS*?

When put into the context of ISO 14001, a gap is the difference between what the standard requires and what existing environmental management practices actually exist in an organization. The gaps analysis compares the existing environmental management practices, methods, and techniques with the requirements of ISO 14001. The results of the study enable management to determine the degree of conformance with the ISO 14001 requirements.

PERFORMING A GAPS ANALYSIS

A gaps analysis is performed by using a matrix of individual ISO 14001 requirements as a baseline. The management of the business, with or without outside help, assesses the current state of affairs relative to each item and records the results. This assessment is not intended to be a complete environmental management system audit; rather, it is an overview of each area of practice to determine what is being done and how that activity meets the ISO 14001 requirements. For each ISO 14001 requirement, the business needs to determine:

- What the standard requires
- What procedures currently exist
- How the current procedures compare with the standard

. If the business has performed an initial review of environmental aspects and impacts, legal and other requirements, this information can be used as input to the gaps analysis process.

The degree of implementation of each of the requirements can be assessed using a scale from 0 to 4. Each point value corresponds to an increasing level of implemented systems that meets the requirements of ISO 14001. The scale is defined as follows:

- 0 = Nothing exists at all
- 1 = A program description exists but is not implemented
- 2 = A system is developed and partially implemented or is fragmented and uncontrolled
- 3 = A system is developed and implemented but does not meet the intent of ISO 14001
- 4 = The system is developed and implemented and meets the ISO 14001 intent

The results of the study should be recorded and used as a basis for determining resource priorities for implementing the environmental management system. Systems having a rating of 0 or 1 obviously will require much more effort than those receiving 3 or 4 ratings. The purpose of this study is to obtain results that can be used to estimate the activities and extent of effort required to implement an ISO 14001 environmental management system, and to provide the basis for allocation of resources. It is not expected that the final outcome will be accurately predicted; rather, the study is a judgment of what will be necessary.

Developing an Action Plan

Once each requirement has been assessed, and the business has an overall picture of its activities relative to the ISO 14001 standard, it can begin to develop action plans for each requirement. The key to successful ISO 14001 implementation is to integrate the planning efforts with the existing planning activities of the business. As with any planning effort, the gaps analysis is used as an input tool to develop specific action items and priorities. Management also needs to ascertain certain dimensions of the ISO 14001 project before it commits to the implementation path. These dimensions include:

- The size of the project (is it a major effort or a minor adjustment?)
- The size of each gap (does the business have to develop and implement any major program components?)
- How can each gap be filled (what exactly will the company be doing?)
- The amount of current information available (what else does the company need to know to do this?)
- The nature and type of approvals required (what choices and options require additional approvals?)
- Deadlines and timing (how does the ISO 14001 effort integrate with the rest of the business's activities?)
- Resources required (who should be involved and what are the available budgets?)

Estimating Project Resources

The resource estimation of the ISO 14001 project may be the most difficult to achieve for the simple reason that it is often hard to predict the efficiency of future implementation activities. Competing priorities, changes in management, marketplace shifts, employee attrition, and changing laws and regulations often have large impacts on how businesses conduct their operations. Long-term projects, such as a major systems implementation, can often change in scope and budget during the course of the project due to factors beyond the project manager's control. Thus, flexibility becomes another important component of the ISO 14001 planning process.

While the recommended sequence of implementation activities is given in Chapter 19, there are some general considerations that can be applied to the resource planing effort. These include:

- **Does the business have any previous experience with similar systems projects, such as ISO 9000?** This type of experience can yield valuable information about resource expenditures for management systems development and implementation.
- **Can the project be applied to a pilot or a small operating group first, and expanded into other areas of the business later?** Smaller projects are easier to estimate and the management systems are easier to define.
- **Can resources be leveraged through trade associations or universities?** Expertise gained through these types of relationships can greatly assist with specific problem solving. They can also provide management with a fresh perspective on problems that may seem unsolvable or particularly vexing.
- **What are the resources required for development and operation?** The initial development and implementation often takes a great deal more effort than is required for the ongoing maintenance and operation of a well-designed system. The project efforts (money and deployment of human resources) will be high initially and reduce after the systems become operational.

Once the gaps analysis is performed and the results are reviewed with senior management, the next logical step is to consider implementing the systems approach to manage and reduce environmental risk. ISO 14000 and the legal and practical techniques addressed in the preceding chapters can be utilized to develop an effective program. In Chapter 19 we discuss how this can be accomplished.

Chapter 19

IMPLEMENTING THE SYSTEMS APPROACH

The systems approach to environmental management is a method to manage and reduce environmental risk. Implementation of such a system cannot begin before there is commitment from top management with accountability for continuing and active participation in promoting and overseeing the environmental management system. The systems approach can be used by managers either to create or to enhance an existing environmental management system.

At the outset, we will review some important concepts regarding how top management should get involved in the development and continuing operation of an environmental management system. Next, we will review the 17 mandatory elements of an ISO 14001 certified environmental management system which provide a framework for managing environmental risks on a daily basis. This chapter will also be supplemented with specific suggestions derived from Chapters 10 to 17 on voluntary strategies to be employed to reduce risks while at the same time effectively managing the system. Included will be examples of good business practices contained in the National Center for Preventive Law's (NCPL) corporate compliance principles project. The NCPL Commission on Corporate Compliance has recommended guidelines for creating an effective corporate compliance program which would necessarily support an environmental management system for those companies that impact the environment.

All of the standards, principles, considerations, and examples contained in this chapter are intended to demonstrate how the systems approach can be used to manage and reduce environmental risk. ISO 14001 provides 17 elements that must be present for the environmental management system to comply with the standard. ISO 14004 and the NCPL Commission, on the other hand, simply make recommendations regarding how companies can use more effective means to achieve the level of results they desire.

COMMITMENT FROM THE TOP

Commitment from senior management to improve the environmental performance of the company is necessary whether a system is being designed pursuant to ISO 14001 or other standards. Without this support, it is unlikely that efforts to implement an effective system will succeed. For this reason, top management needs to recognize that environmental performance is among the highest corporate priorities and act accordingly. The system that is created should reflect the business's culture, ethos, and corporate objectives (NCPL, 1996).

The NCPL Commission found that the highest governing authority within the business should endorse the program and give it visible support. Many environmental policies begin with a personal statement from the chief executive officer, the board of directors, or senior management regarding the company's commitment to the environment. For example, David Simon, Group Chief Executive of British Petroleum, introduced BP's policy in March 1993 with his personal commitment to BP's health, safety, and environment policy. As a company, Johnson & Johnson, introduced its new environmental leadership policy with a statement by Chairman and CEO Ralph Larsen to the shareholders on April 25, 1991. American Express began its *Environmental Principles* with a statement from the Public Responsibility Committee of the company's board of directors regarding how the committee will oversee management's commitment to environmental policies and practices.

Significant input, commitment, and leadership from top management is necessary to make an environmental management system successful. Management can demonstrate their interest and commitment to the environmental management system by reviewing and approving specific provisions of the program, making clear assignments of responsibility for the program, and holding themselves accountable for those compliance activities that they initiate or oversee (NCPL, 1996).

Top management needs to do more than just set an example for the rest of the company to follow in protecting the environment and minimizing environmental impacts. The environmental management system should be designed with input from knowledgeable people throughout the business and from outside consultants if necessary. Open lines of communication with internal and external stakeholders can be established and maintained during this process so that the system is effectively created and received. Thoughtful environmental planning combined with product and process life-cycle analysis can increase the likelihood of achieving environmental performance targets as employees learn to manage and reduce their risks and liabilities.

Not all companies will have the resources to create an effective environmental management system. The program needs to be tailored and fine tuned with specific regard to the size, form, complexity, and history of the business. Some companies' systems will be more formal than others. All should be in writing and set forth program definitions and operating practices in clear terms. The environmental management system should be readily available to all employees. Employees should easily understand environmental management system procedures and where to get help or questions answered. To assure fundamental fairness in the operation of the system, all employees subject to the system should be treated equally and consistently. Mechanisms should be written into the system to prevent retaliation for raising compliance issues.

Both ISO 14001 and the NCPL leave the choice of the format of the system up to the company itself. The company will set the level of compliance activity based upon the company's existing policy, the totality of the company's environmental impacts, the size of the company, the type of business it conducts, its available assets, and its existing reputation and external relations.

HOW TO GET STARTED

Chapter 4 addresses the necessary steps for a company to proceed with an ISO 14001 initial environmental review. The purpose of the review is to identify legislative and regulatory requirements; to identify the environmental aspects of its activities, products, or services; and to ascertain which aspects create significant environmental impacts, risks, and liabilities. This preliminary review should also be used to evaluate existing environmental performance. Relevant internal criteria, external standards, codes of practice, and sets of principles and guidelines can be collected to benchmark existing performance. The review should also consider existing policies and procedures that deal with other related activities, such as procurement and contracts. If there are any previous incidents of nonconformance, the review can include feedback from prior investigations. The initial review can identify what functions or activities of the company's other systems enable or impede the company's environmental performance. Collecting and analyzing this information during the initial review can also serve the dual purpose of discovering opportunities for competitive advantage and a baseline of existing views of stakeholders.

ISO 14004 recommends that information be collected by submitting questionnaires to management, employees, and stakeholders; conducting interviews with key personnel; inspecting and measuring existing systems; reviewing pertinent records; and benchmarking. Government agencies can be contacted to identify compliance permits. Local or regional libraries or databases can be queried, along with industry associations, larger customer organizations, manufacturers of equipment in use, and other related businesses. At the completion of the initial review, the company is ready to begin drafting an environmental policy.

1. Environmental Policy

A corporate environmental policy will establish an overall sense of direction and set the principles of action for the business. The goals of environmental responsibility and performance required of the business will be judged against all subsequent actions. Chapter 10 provides a review of the contents of many environmental policies that were developed prior to the release of ISO 14001. The following section incorporates many of these concepts with NCPL considerations and examples of good corporate compliance practices. The following factors should be included or considered in the development of a sound environmental policy:

The Mission, Vision, Core Values, and Beliefs of the Business:

Many companies, like Hewlett-Packard, state that they are committed to conducting their businesses in an ethical and socially responsible manner. HP goes further than most by acknowledging that it has an "aggressive approach to environmental management," which also includes occupational health, industrial hygiene, safety management, and ecological protection. HP believes its approach is "consistent with the spirit and intent of our established corporate objectives and cultural values."

ComEd begins its policy with the statement that it has long been committed to protecting the environment because of its special position as the major supplier of electrical energy in Northern Illinois. It adds, like many other companies, a statement that it believes "our customers want and expect us to respect the environment." Further, ComEd makes clear that the company is a member of each community in which it operates. Thus, its employees are deeply interested in the effects its operations have on the environment and the health and safety in the service area. Like HP, ComEd states that it is taking "aggressive action" to protect the environment, and adds that it takes pride in its compliance record and cooperation with authorities. The tone of the HP and ComEd policies immediately indicates that their policies reflect, incorporate, and integrate with their business's culture, ethos, and corporate objectives.

Environmental Responsibility and Leadership:

Two closely related topics are environmental responsibility and leadership. To indicate responsibility, Bristol-Myers Squibb says it "will ensure that each employee understands the importance of, and is responsible and accountable for, integrating environmental health and safety considerations into their daily responsibilities." Johnson & Johnson states that its goal is environmental leadership. "We're committed to it for two reasons. One is our value system as reflected in Our Credo. The second is our belief that the public's concern about the environment is a very natural extension of each individual's concern about his or her own personal health and the health of his or her family." Both of these statements also underscore an attempt by the businesses to demonstrate that environmental leadership and responsibility are important corporate objectives.

Sustainable Development:

There is a growing trend for companies to endorse the concept of "sustainable development" in their policy statements. Sustainability simply means leaving future generations with a better world economically, socially, and ecologically (Spedding, 1996). Xerox states that it has an aggressive environmental management program that creates products for a sustainable future. It cites, as one example, the introduction of its new Document Centre® products which are helping to create a fully digitized workplace where paper is just one mode of information transmission. Combined with corporate environmental initiatives, Xerox is designing products that are manufactured, used, and reused in a manner that improves the environment. A significant transition is occurring — from the old mode of business to electronic or networked offices and multifunction products with multiple life cycles. Xerox believes this change is vital for future sustainability.

Baxter has another view on sustainability. Under that heading Baxter states "we will strive to conserve natural resources and minimize or eliminate adverse environmental effects and risks associated with our products, services and operations."

Communications with Interested Parties:

Virtually every company that we reviewed recognizes the need for open communications with interested parties. Environmental communications expert A.J. Grant points out that effective communications with external stakeholders cannot and should not begin before a company understands the message and delivers it effectively and meaningfully to internal stakeholders. The message must begin with an environmental policy that is soundly developed and available to all employees. Employees need to develop trust in management that the policy is meant to be implemented in a consistent, fair, and effective manner. At that point the policy can be made public.

Union Carbide's policy contains a commitment to establish and present a responsible and consistent position to government and the public on health, safety, and environmental matters affecting its products and operations. General Motors confirms that it will continue to participate actively in educating the public regarding environmental conservation. In an effort to improve dialogue and community relations, Chevron's Oak Point Plant organized a community advisory panel composed of civic leaders and members of the community to discuss environmental issues and concerns. All three of these companies have taken affirmative steps to work with the communities in which their plants are located to encourage open communications regarding operation and safety risks.

Continual Improvement:

Only the most recent policies contain a commitment to continual improvement, an ISO 9000 and 14000 required term that picks up the concept from the Japanese term *kaizan*, which means small, ongoing, and continuous improvements (Mcinerney and White, 1995).

Xerox states that it is dedicated to the concept of continuous improvement of its performance in environment, health, and safety. The Public Service Company of Colorado similarly states that it is continually searching for ways to improve its environmental and safety performance. Duke Power requires each employee to pledge to "continually look for ways to improve performance and better protect the environment." StorageTek states that it is striving for continuous improvement in pollution prevention, waste minimization, and resource conservation.

Pollution Prevention:

Prevention of pollution is a key topic contained in all of the policies we checked, and a required element of policy conforming to ISO 14001. For example, Monsanto has pledged to reduce all toxic and hazardous releases and emissions, working toward an ultimate goal of zero effect. 3M states that it will prevent pollution at the source whenever and wherever possible. DuPont declares that it will "drive towards the generation of zero waste at the source." Dupont further commits to reuse and recycle materials to minimize the need for treatment or disposal and to conserve resources. Where waste is generated, it will be handled and disposed of safely and responsibly.

Johnson & Johnson has an impressive record of pollution prevention from 1987 to 1989. It decreased by 50% reportable chemical releases from its U.S. manufacturing facilities. The company has eliminated ozone depleting chlorofluorocarbons (CFCs) from more than 120 products and processes worldwide and upgraded 1000 of its storage tanks to state of the art standards in 1993.

Coordination with Other Organizational Policies (e.g., Quality, Occupational Health, and Safety):

Thoughtful companies integrate their environmental management system with business operations and other company compliance policies. Lockheed Martin, for example, has agreed to integrate its environmental, safety, and health considerations into strategic business discussions, engineering design, procurement, facilities management, and production. It seeks opportunities to improve safety, health, and the environment; enhance competition; and reduce environmental, safety, and health costs.

The environmental management system should be consistent with all other existing company policies. Each of the company's multiple compliance policies will typically identify important business objectives and goals. To ensure consistency, the environmental program can specifically incorporate portions of existing policies. The NCPL Commission describes a company, for example, that incorporated its existing code of ethics, vision statement and corporate guidelines, as well as existing compliance activities, as part of its overall compliance program. An environmental management system is an integral part of a company's overall efforts to achieve compliance and reduce risks and liabilities.

Specific Local or Regional Conditions:

Many companies' operations have specific impacts on local or regional conditions. Some of these impacts can be positive. As mentioned in Chapter 17, BASF is co-locating plants so that the industrial waste created in one can be used as a fuel in another. This action stimulates the economy, improves the environment, and avoids the high cost of transporting and disposing of industrial wastes. Other companies have designed transportation programs including ride-share, van pools and other transportation alternatives, and have made other accommodations to employees including telecommuting in order to reduce environmental impacts. Some of these companies provide incentives for employees to find alternative means of transportation to and from the workplace. Local or regional planning initiatives can be incorporated into the environmental policy of the company to reduce impacts on ecosystems.

Compliance with Relevant Environmental Regulations, Laws, and Other Criteria to which the Business Subscribes:

Virtually all of the companies commit to compliance with relevant laws. This element is required by ISO 14001. As an example, BFI states that it will comply with all applicable environmental health and safety laws and regulations. These actions are intended to minimize adverse environmental health or safety effects from the company's business activities to achieve a cleaner global environment.

Minimize any Significant Adverse Environmental Impacts of New Developments through the Use of the Integrated Environmental Management Procedures and Planning:

A successful management system can be tailored to the needs of a growing business that is experiencing increased regulatory responsibilities. While business is expanding, companies need to avoid developing environmental management systems out of proportion to a company's risks of noncompliance. Companies must focus on functions to achieve a proper balance. In the worker safety area, for example, the NCPL Commission found a risk assessment of operating activities might be completed to identify high-risk operations. These operations could then be addressed with greater attention and detail. Lower-risk operations, on the other hand, might be addressed with less comprehensive direction and monitoring. ISO 14001 requires a periodic review to realign resources and risks.

Development of Environmental Performance Evaluation Procedures and Associated Indicators:

A company needs to develop methods to assure itself that the environmental management system is working. Both the EPA and the states have been seeking to discover appropriate environmental indicators to let them know that their regulatory programs are achieving the greatest benefits for the minimum costs. A company may consider the benefits of participation in voluntary programs, such as the EPA's Project XL, that seek to embody a systematic approach to environmental protection and test alternatives to regulatory requirements. Similarly, a company needs to have indicators, such as a lack of environmental compliance violations or public interest group complaints, a continuous record of reduced waste and more recycling, or other measures to let it know that the system is working and objectives are being achieved.

Embody Life-Cycle Thinking and Product Redesign:

Conoco states in its environmental policy that it encourages life-cycle assessments in the development of its products. Chapter 17 is replete with examples of other companies that incorporate this thinking into daily business operations. These examples include redesign of products like vacuum cleaners, cars, pesticides, herbicides, tires, and copiers to reduce their environmental impacts. Many companies have undertaken wastestream analyses that have changed production methods. These studies have also resulted in product redesign to eliminate waste, as well as the use of toxic substances and unnecessary parts, so that products can have a second or third useful life and not be discarded into landfills until their usefulness is truly exhausted. Environmental policies that endorse life-cycle analyses, coupled with sustainable development and continuous improvement, are on the cutting edge of environmental proactive thinking.

Reduce Waste and Consumption of Resources (Materials, Fuel, Energy), and Commit to Recovery and Recycling, as Opposed to Disposal Where Feasible:

Commercial and residential carpet tiles, refrigerators, VCRs, and CRT glass from computer and television monitors are just a few examples of consumer goods that can be recycled to avoid wasting resources.

Johnson & Johnson's recycling efforts have saved the equivalent of 370,000 trees each year. Since 1972 its energy conservation program has saved the equivalent of 32 million gallons of oil. Since 1988 the company has eliminated more than 5.5 million pounds of packaging from 32 products.

Xerox has implemented programs that enable its customers to return supplies to Xerox for reuse and recycling, such as copy and print cartridges and toner containers.

Recycling is important; however, elimination is better because corporate and natural resources are spared when the waste is never created in the first place. To improve their competitive position, not incrementally but radically, forward-looking companies from around the world are embracing the ultimate environmental goal: zero emission (Mcinerney and White, 1995).

Management of Environmentally Risky Products:

Consistent with the "zero waste equals zero defects" philosophy of Mcinerney and White, Chevron has eliminated products, such as agricultural chemicals and paint thinners, where the costs to properly manage the products' environmental risks and meet regulatory requirements outweighed their earnings potential. Monsanto has bioengineered plants to reduce the use of pesticides and to make herbicides more effective and less harmful to the environment.

Education and Training:

An environmental policy can include a commitment to communicate appropriate environmental impacts information to all of the business's employees and provide them with the necessary information and skills to deal with the environmental impacts issues and risks that each employee may encounter. The NCPL Commission noted that some companies develop a legal-risk analysis to help train their employees on communications strategies. This involves determining the business's activities, evaluating and prioritizing the type of legal risks encountered during those activities, and developing an appropriate communications program to manage and minimize risks (NCPL, 1996).

Sharing Environmental Experience:

British Petroleum has committed to communicate openly with those who live or work in the vicinity of its facilities to ensure their understanding of BP's operations, and BP's understanding of their concerns. Concord Resources has committed to encourage open dialogue with the public to assist in their understanding of the environmental impacts of Concord's activities. Baltimore Gas and Electric has agreed to respond to its customers, neighbors, employees, regulators, and others whenever they raise concerns about the environmental impacts of its business. Sun Microsystems shares its program successes with employees, customers, and the general public to further the efforts of environmental stewardship.

Encourage the Use of an Environmental Management System by Suppliers and Contractors:

Baxter's environmental policy includes a commitment to work with its suppliers and contractors to enhance environmental performance. Bristol-Myers Squibb goes further by stating that when feasible, it will give preference to suppliers and contractors whose environmental health and safety commitment and practices are consistent with its own, and who have demonstrated environmentally responsible products, services, and management.

There are compelling legal reasons for this management commitment. The NCPL Commission said that to ensure accountability mechanisms apply to all sources of liability and legal risk, a company may wish to extend some aspects of its management system practices outside of its organization. For example, some actions of external corporate agents or independent contractors can create significant liability for the company. Accordingly, a business has a strong interest in holding these individuals accountable for their environmental impacts. Companies may wish to incorporate provisions in related contracts requiring conformance by agents and contractors acting for the firms and providing for reporting and reviews concerning conformance by these outside parties (NCPL, 1996).

The Commission also found that a company may wish to inform outside parties, with which it conducts business, of the company's environmental expectations. For example, a company may want to consider requiring specific compliance results in agreements or contracts with agents and vendors. As a part of its environmental management system, a company may require that its agents and vendors be ISO 14001 certified. This requirement can be included in the written contracts that the company negotiates and may also be a part of a vendor certification process.

While the environmental policy is perhaps the centerpiece of an environmental management system, it is still only one component of a fully integrated plan to minimize environmental risks and reduce impacts. The remaining sixteen ISO 14001 elements round out an effective environmental management system.

2. Environmental Aspects

ISO 14001 recognizes that a business's "policy, objectives and targets should be based on knowledge about environmental aspects and significant environmental impacts associated with its activities, products or services. This can ensure that significant environmental impacts associated with these aspects are taken into account in setting environmental objectives." In order to manage and reduce environmental risks and liabilities, a company needs to know the environmental aspects of its activities, products, and services.

The most obvious environmental impact of a company is its wastestreams. Chapter 17 has examples of companies that studied their wastestreams and found more than they expected. Elimination of wastestreams directly cuts down on environmental aspects and impacts. Other impacts may include how products are packaged, how energy sources are used, and how products are manufactured — to name a few. Companies can have energy audits conducted to eliminate unnecessary uses of energy and natural resources. To further reduce impacts, companies may require each employee to develop a personal environmental impacts plan. Companies also need to establish and maintain separate procedures to evaluate the impacts of their operations on the environment.

Some environmental impacts are not so obvious. But unless they are identified, they can cause liability for the company. The NCPL Commission found that in order to contain risks, companies need to identify nonobvious and incipient misconduct that tends to promote illegal actions (NCPL, 1996). For example, if a company has a waste disposal policy that requires employees to use haulers that charge the lowest disposal costs possible, illegal discharge of the company's waste may result. Similarly, if corporate environmental goals are unrealistic, employees may be tempted to lie in reports in order to appear to achieve those goals. A company has to be prepared for a full range of collateral, unwanted consequences if it sets the pollution bar too high.

Changes in operations, products, or services may affect the environmental aspects and their associated impacts. A management system must be responsive to daily operations and be dynamic to account for changes in business activities (NCPL, 1996). The system must be appreciated by those faced with new and challenging environmental issues. Employees learn something new every day and face uncertainty in solving complex problems presented by the workplace. Responsive environmental management systems have built-in components to instill confidence in employees and to motivate them in the face of uncertainty. One company explained its program to employees by illustrating its intended implementation with concrete examples drawn from the business's specific work-related activities. These illustrations can take into consideration changes that can occur in the workplace as well as failures and how to responsibly deal with them.

Before a process failure occurs, companies should have plans in place to deal with the consequences both internally and externally. DuPont has agreed to inform its employees and the public about the safety and health effects of its products and workplace chemicals and to provide leadership in establishing programs to respond to emergencies involving hazardous materials in communities where the company has a significant presence. Conoco has agreed to maintain emergency preparedness plan and response capabilities.

In addition to developing hazard awareness and crisis management plans, companies need to find out how frequently the situation may arise that could lead to the impact. This will determine the level of effort required to minimize the risk of occurrence of an incident.

To understand the degree of risk, ISO 14001 requires the following environmental concerns need to be addressed: the scale of the impact, the severity of the impact, the probability of occurrence, and the duration of impact. Once those concerns have been considered, ISO 14001 requires that the following business concerns need to be addressed as well: the potential regulatory and legal exposure, the difficulty of changing the impact, the cost of changing the impact, the effect of change on other activities and processes, the concerns of interested parties, and the effect on the public image of the organization.

Environmental management systems will be most useful when they are carefully tailored to fit each business's unique situation. The same may be true regarding operating units within a business. Each different operating unit of a business may need its own environmental management system. The system must be adapted to address multiple different and possibly inconsistent legal requirements. For example, a company with global operations must confront many different, and potentially conflicting, legal and environmental requirements. One international company adopted a core of global standards, but permitted local management to modify portions of its program to take into account local needs and requirements (NCPL, 1996).

3. Legal and Other Requirements

A formal procedure is required to identify ongoing legal and other requirements that are applicable to the environmental aspects of the business. The standard requires identification of not only governmental regulations, but also of industry associations or groups and commercial standards of practice and professional codes. Under the systems approach to the minimization of risk, the procedure should also identify the range of possible consequences of a business's actions, including civil and regulatory violations. Internal controls procedures can then be implemented to effectively eliminate noncompliance. The identified requirements must be readily accessible, updated, and available to employees.

For compliance purposes a business needs to identify regulatory requirements that are specific to its activities (e.g., site operating permits), to its products or services, and to its industry. The business also needs to identify relevant general environmental laws, authorizations, licenses, and permits. To keep track of all relevant legal requirements, a business can establish and maintain a list of all laws and regulations pertaining to its activities, products, or services in electronic or hard-copy format and a procedure for identifying changes that occur to legal and other regulatory requirements.

The NCPL Commission found that in creating a compliance program a business may thoroughly examine its liability-risk profile. To accomplish this profiling, it may be useful to examine the risk experience of the business's industry. For example, a company can have its attorney prepare a report on compliance problems that the company's industry has experienced. To complete a liability-risk profile, a company can assess its own past compliance history. One company assigned an in-house attorney to conduct both a search of the company's files and a compliance audit to determine compliance risks that the company had faced. While the involvement of an attorney may help preserve the attorney-client privilege, to the extent that such studies are viewed as management tools or are disclosed to public agencies to demonstrate compliance diligence, these audits can be subject to disclosure in civil and criminal cases (NCPL, 1996).

Management systems should help companies focus upon risks that a business most frequently confronts. A listing of recently encountered environmental risks and problems can be useful in ensuring that a complete set of corresponding compliance program elements is adopted. Companies that operate within a broad regulatory framework can begin their risk assessment by examining their own and other like companies' histories of violations and citations. Many companies also conduct "litigation audits" as a starting point for assessing their legal risks (NCPL, 1996).

4. Objectives and Targets

ISO 14001 requires that there be documented environmental objectives and targets set at each relevant function and level within the business. Companies need to set forth these goals and methods for achieving them in a clear and straightforward manner (NCPL, 1996). Objectives are the overall goals for environmental performance that are identified in the policy statement. When establishing its objectives, a business should take into account the relevant findings from the initial review and the identified environmental aspects and associated impacts. ISO 14001 provides that "[e]nvironmental targets can be set to achieve their objectives within a specific time frame. The targets should be specific and measurable. When the objectives and targets are set, the organization should consider establishing measurable environmental performance indicators . . . Objectives and targets can apply broadly across an organization or more narrowly to site-specific or individual activities."

The environmental objectives and targets need to reflect both the goals of the environmental policy and significant environmental impacts associated with the business, activities, products, or services. In setting objectives and targets, management has to consider whether the employees responsible for achieving the objectives and targets have had sufficient input into their development and whether the views of interested parties have been adequately considered.

The following are some examples of companies who have made public commitments to achieving environmental objectives:

Reduce waste and the depletion of resources:

In 1991 Hitachi was the first Japanese company to set industrial waste reduction goals for the year 2000 (Mcinerney and White, 1995).

Reduce or eliminate the release of pollutants into the environment:

Westinghouse promised to reduce toxic chemicals released to the air by 50% by 1995 and 90% by 2000, based on 1988 levels.

Design products to minimize their environmental impact in production, use, and disposal:

Many companies have committed to designing products for disassembly so that they can be recycled into new products. Takeback laws in Japan and Germany have provided financial and ecological incentives to companies to redesign products ranging from vacuum cleaners to BMW sports cars to reduce environmental impacts. The EPA has signed partnership agreements with office equipment manufacturers representing approximately 90% of the office equipment market to produce computers, monitors, printers, and facsimile machines that automatically power down when not being used and to manufacture copiers that automatically make double-sided copies and turn off when not being used.

Control the environmental impact of sources of raw material:

WMX Technologies has committed to using renewable natural resources, such as water, soils, and forests, in a sustainable manner and to offering services to make degraded resources once again usable. WMX has also agreed to conserve nonrenewable natural resources through efficient use and careful planning.

Minimize any significant adverse environmental impact of new developments:

Chevron has committed to work closely with local tribes and government officials in New Guinea to ensure that the cultural and environmental integrity of the region is retained. Chevron has buried pipelines to minimize rain forest damage and helped to build schools and clinics.

Sun Company has committed to pay special attention to the protection of the surrounding environment at present facilities and when planning for new facilities or operations.

Westinghouse has committed to minimizing adverse environmental impacts in the planning and development stage of new infrastructure. Its Pelican Bay Project near Naples, FL combines environmental planning and business success, and it has been lauded by environmentalists and honored by the state for these efforts.

Promote environmental awareness among employees and the community:

Chevron's Citizen's Advisory Group at its Oak Point Chemical Plant is composed of civic leaders and concerned members of the community who are reviewing the company's plans to handle emergency incidents. Chevron also conducts and periodically updates training programs to help employees understand how social, political, and legal aspects of society's environmental values affect its business. The training program emphasizes the company's dedication to environmental compliance and risk management.

Howe Sound Pulp and Paper Company held public meetings in Canada with slide shows and videos to show its commitment to minimize environmental impacts in Canada's forestry industry. (Mcinerney and White, 1995).

The management of Inter-Continental Hotels held meetings with its employees to demonstrate its commitment to reduce its environmental impacts. Skeptical employees were won over after a time and subsequent meetings produced small but tangible benefits by reducing impacts and greening the hotel system (Mcinerney and White, 1995).

Environmental performance indicators can be used to measure progress towards an objective. Examples of performance indicators include the following:

Quantity of raw material or energy used and the quantity of emissions such as CO_2:

Nissan has recognized that the long-term viability of its business may depend on developing alternatives to the internal combustion engine. Nissan is studying those alternatives and its ability as a huge automobile manufacturer to make a significant contribution to the reduction of conventional pollutants such as CO_2 (Mcinerney and White, 1995).

Sun Company is measuring the reduction of the use of water in its facilities in an effort to decrease the quantity of its water discharge streams.

From 1987 to 1989, the company cut in half reportable chemical releases from its U.S. manufacturing facilities. It eliminated ozone-depleting chlorofluorocarbons (CFCs) from more than 120 products and processes worldwide and is committed to total CFC replacement.

Union Carbide reduced chemical air emissions at all but one location worldwide. Potential exposure levels in the community were found to be at least 1000 times lower than workplace standards.

Monsanto Company pledged to reduce all toxic and hazardous releases and emissions, working toward an ultimate goal of zero effect.

Percentage of material recycled and used in packaging:

Coors measures the amount of waste it has recycled, including 107 tons of office and colored paper; 7703 tons of corrugated paperboard; 19,568 wooden pallets repaired; 16,965 cubic yards of wooden pallets composted; 1,087,600 gallons of process sludge composted; 125 tons of stretch wrap; and 89,000 pounds of plastic banding recycled.

Johnson & Johnson has calculated that its current U.S. paper-recycling efforts alone are saving the equivalent of 370,000 trees each year. Since mid-1988, the company has eliminated more than 5.5 million pounds of packaging from 32 products.

Union Carbide's efforts to collect paper, plastic, metal, and used oil at U.S. sites yielded 17 million pounds of recyclables in 1992. The 1993 total was expected to be 19 million pounds. Paper recycling, including cardboard and newsprint, totaled 1.5 million pounds in 1992. The 1993 total was projected to be 2 million pounds. These efforts saved 17,000 50-foot trees; 7 million gallons of water; 2500 barrels of oil; and 3000 cubic yards of landfill space.

Xerox uses permanent parts in its shipping containers and carts in manufacturing operations; these were returned to suppliers for refilling and reuse. Xerox also uses recycled-content corrugated cartons for shipping consumable supplies.

5. Environmental Management Program

The ISO 14001 standard requires that a program or programs be established and maintained for achieving the identified objectives and targets. The program to achieve good environmental results is only a part of the larger environmental management system designed to manage and reduce environmental risks, impacts, and liabilities. The program must define the means, schedule, and responsibility at each relevant function and level to achieve the objectives and targets. The program should also be amended as necessary to include new developments or modified activities and products or services, yet still attain the objectives and targets.

The NCPL Commission found that a written report of how the compliance program was created and implemented can be very useful. The environmental management program similarly needs to have sufficient detail for an employee to clearly understand how the program was first developed and whether the planning process involved all responsible parties. The environmental management program should document the specific steps necessary for the business to succeed in achieving objectives and targets and who is responsible for achieving those goals.

A written record can also be prepared to demonstrate how the system operates on a daily basis and to monitor the system's current effectiveness (NCPL, 1996). A written record can also be used to defend the program's effectiveness and tailor the program for more effective use in the future. The environmental management program also should have a process built in for periodic reviews of the program.

The NCPL Commission found that an effective compliance program is a dynamic process that can be designed to be flexible and modified, when appropriate, to reflect changed circumstances. It must be able to respond to new business activities or other changes in the business. The NCPL Commission further found that a successful program is usually adaptive to changes in the company's environment and can include components that respond to the new plans and conditions of the business. Unplanned changes can produce new compliance risks, so an effective program should be able to adapt to meet those risks without compromising the system's integrity.

The following are some examples of how businesses maintain their programs in the face of changing circumstances:

Conoco's environmental program dates back to February 29, 1968. To adjust the program to changing conditions, Conoco regularly conducts environmental quality assurance audits as one way to determine if it is achieving its goal of conducting business with "respect and care for the environments in which [Conoco] operates."

Sun Company, Inc. changed the direction of its environmental program in 1993 when it became the first Fortune 500 company to endorse the *CERES Principles* in order to demonstrate its commitment to public accountability for environmental protection. Sun publishes an annual report on its performance and accomplishments of its health, environmental, and safety program and reports on its progress in consistent, measurable terms. To promote public accountability among others in the industrial community, Sun cooperates with the environmental and social investment communities by completing an annual CERES report.

IBM conducts audits and self-assessments of its compliance with its environmental policy, measures progress of its environmental affairs performance, and reports periodically to its Board of Directors.

Johnson & Johnson instituted a periodic environmental assessment program by independent outside reviewers. More than 200 of its worldwide facilities are covered by these third-party audits. When no standards are specified, or in countries where environmental standards are less stringent than in the U.S., Johnson & Johnson follows its higher corporate requirements and guidelines.

6. Structure and Responsibility

ISO 14001 requires environmental roles, responsibilities, and authorities to be defined, documented, and communicated to all relevant employees. Financial, technological, and human resources essential to the environmental management system must also be provided. A management representative must be selected who reports directly to senior management on environmental management system performance and improvement. The management representative must ensure that the environmental management system is established, implemented, and maintained.

It is important for the individual who is selected as management representative to take ownership of the environmental management system and to exert overall responsibility for initiating, coordinating, and monitoring the environmental management system efforts (NCPL, 1996). This individual obviously has to have the necessary degree of clout and respect within the business to make the system work. A high level of authority, coupled with access to senior management, assures that the management representative and the environmental management system are perceived as important activities by the other people in the business. The individual must also have the right combination of personal characteristics and knowledge of systems operations to be effective in leading and promoting environmental performance improvements (NCPL, 1996). Businesses with regulatory responsibility need to have a carefully defined program that weeds out managers and employees who disregard the law, regardless of the consequences to business operations.

Large businesses need to be careful to allocate sufficient resources to meet the goals of the environmental management system. The goals are broader than merely the objectives and targets of the environmental management program. They include the need for the business to minimize environmental risks and liabilities, as well as other goals such as promoting effective communications, training, and improving industry standards. The management representative is responsible for ensuring that the company finds the correct level of financial contribution to the system to achieve these goals.

ISO 14004 suggests that businesses with limited resources can team with larger companies to share technology and compliance information. They can also create opportunities for similarly situated companies that are interested in developing environmental management systems to define and address common issues, to share knowledge, to facilitate technical development, and to use facilities and consultants jointly to develop environmental management systems. To save expenses, smaller businesses can participate in training and awareness programs conducted by standardization organizations, consultants, associations and chambers of commerce, as well as universities and other environmental research centers.

In order to have a successful environmental management system, participation in and responsibility for the system should not be limited to the management representative. Environmental responsibility needs to be shared throughout the business. Senior management should ensure that incentives for making the system work are uniform, consistent, and spread throughout the business. Incentives and disincentives are important tools for promoting awareness of environmental responsibility and achieving the goals of the environmental management system.

The NCPL Commission found that corporate managers can influence behavior by linking employment treatment with each employee's environmental performance; for example, increased compensation and advancement to employees' furtherance of compliance goals and objectives. Employees throughout the business can be told that it is the policy to allocate incentives and disincentives in accordance with each employee's environmental performance. Corporate leaders can also underscore this message to ensure that the rewards and discipline are in accordance with the relative levels of effort (NCPL, 1996).

Environmental responsibility includes the development of a risk reduction litigation management plan which can include litigation awareness programs and other techniques to avoid or at least minimize liability and legal costs. The management representative can coordinate the development of such a strategy with legal counsel who specialize in risk reduction techniques and identification of liability-causing conduct. The plan should identify the company's overall goals and objectives for the litigation and create a cost-benefit analysis. The plan should address concerns about adverse publicity, as well as negative impacts and possible benefits of litigation. The plan should also include steps the company needs to take if the government commences an investigation; who should retain experienced in-house or outside counsel; how, when, and who should be responsible for analyzing key facts; and who should retain experts and when should they be retained. Companies need to know their corporate history and insurance profile to be able to respond quickly and efficiently to any type of environmental claim whether it comes from the government or a third party, and to be able to notify carriers promptly. Insurance needs have to be addressed on a continuing basis and analyzed according to risk history.

The management representative should have a plan in place to determine how any type of case should be staffed and handled, either internally or using outside counsel when appropriate. The management representative can also be in charge of developing opportunities for employees to learn alternative dispute resolution methods and techniques to promote understanding that such techniques can be used to minimize litigation costs and provide better resolution of disputes.

A risk reduction-litigation management plan should include the following questions in determining whether ADR may be appropriate to resolve the dispute:

- Are there present or foreseeable difficulties in the negotiation which will require time or resources to overcome in order to reach settlement?
- Is the case negotiable, i.e., no precedent-setting issues are involved?
- Is there enough case information to substantiate the violations or claims?
- Is there sufficient time to negotiate in light of court or statutory deadlines, or are the parties to sign a tolling agreement (an understanding that a statutory deadline for starting a lawsuit will be extended)?
- Are all the stakeholders identified? Can they all be persuaded to come to the table?
- Is sufficient authority present? Is each relevant interest group adequately represented?
- Is the table level? Can it be made level?
- Is there a climate of trust and willingness to negotiate – or a way to get there?
- Do the parties agree on the scope of issues to be negotiated or are they willing to allocate time and other resources to coming to agreement?
- Will it be possible to bring the best available information and expertise to the process?

- Can fundamentally different values and assumptions be identified and discussed?
- Will the group be able to identify a sufficient number of legal and economically feasible solutions?
- Are the parties committed to the process?
- Are they prepared to analyze costs and benefits through joint problem-solving?
- Are they prepared to sign a written agreement?
- Are they capable of implementing possible solutions?

There is no one right way, no one right process, for efficient and effective ADR. Flexibility in approach, and familiarity with a range of dispute management tools and approaches, will permit the management representative to most effectively manage the company's environmental disputes.

7. Training, Awareness, and Competence

For ISO 14001, a conformance training and awareness program must be instituted to identify training needs and to ensure that each employee whose work has the potential to create a significant environmental impact receives the appropriate training. ISO 14001 provides that training procedures can be instituted to make employees aware of the importance of conformance to the environmental policy; procedures and the requirements of the environmental management system; significant environmental impacts of their work activities; environmental benefits of improved personal performance; roles and responsibilities in achieving conformance to the policy, procedures, and requirements of the environmental management system; and the potential consequences of departure from specified operating procedures. Employees should be trained at all levels to impress on them the importance of disclosure of possible violations, as well as the procedures to do so. Legal counsel can assist companies with the development of litigation awareness programs to identify liability-causing conduct. Typical training programs include a variety of topics related to legal requirements, company values, and the means to consider these in company business decisions and actions (NCPL, 1996). Environmental training goes beyond aspects of health and safety by requiring proven competency. Personnel whose tasks can cause significant environmental impacts must be competent based on education, training, and experience. Proper training will require employees to be responsible and accountable for their business's environmental impacts.

Education and training programs are important to ensure that employees have appropriate and current knowledge of regulatory requirements, internal standards, and the business's policies and objectives. The level and detail of training, however, may vary according to the tasks performed by employees. It is important to match training to the tasks routinely handled by the employees.

Training programs need to be designed so that employee training needs are identified and plans developed to address their specific needs. The NCPL Commission suggests that an effective training program can be targeted to reach an intended audience and should be understandable, accessible, and practical (NCPL, 1996). Interactive computer training programs and software packages have been developed by innovative companies like Compliance Systems Legal Group® in Warwick, RI which have the capability of providing many employees with effective training. The employees' performance is monitored and results tabulated for review and evaluation purposes. Formal documentation of the type of training given and the results of the training exercises is useful to keep a record of compliance activities and to measure the effectiveness of the training received (NCPL, 1996).

8. Communications

As set forth in Chapter 7, communications can include establishing processes to report internally and externally on the environmental activities and issues to demonstrate management's commitment to improving environmental performance. The company also has to deal with concerns and questions raised about the environmental aspects of the business's activities, products, or services. At the same time, communications need to raise awareness among stakeholders of the business's environmental policies, objectives, targets, and programs.

The NCPL Commission found that a good business practice is to have a communications component for the compliance program that makes employees and other agents aware of the applicable standards of conduct and to promote compliance (NCPL, 1996). The formality of the communications component is likely to increase with the size of the business.

ISO 14001 requires that a company set up and maintain procedures for internal communications between the various levels and functions of the business. A process needs to be created for receiving, documenting, and responding to employees' and other interested parties' concerns.

The business also needs to have a documented plan for communicating the business's environmental policy and performance. A plan to disseminate the results of environmental management system audits and reviews to appropriate people in the business is also necessary. Protection of the audit results from unauthorized personnel and third parties is important, along with keeping the results confidential and privileged where appropriate.

Public environmental reporting is becoming a popular way for companies to distribute their environmental results. Environmental reports, as discussed in Chapters 2 and 10, can include the business profile and the text of the environmental policy, objectives, and targets. Beyond this, ISO 14004 recommends that the public report can also include environmental management processes (including interested-party involvement and employee recognition); environmental performance evaluations (including releases, resource conservation, compliance, product stewardship, and risk); opportunities for improvement; supplementary information (including glossaries); and independent verification of the contents. The level of detail contained in these public reports should be directly related to the degree of exposure management wants to give its environmental management system. Companies with relatively new systems should examine this question carefully and consider alternatives to public reporting until the effectiveness of the company's program is proven and the company is ready to discuss openly its environmental record and future plans.

Internal and external environmental communications and reporting involve two-way communications which require the dissemination of understandable and adequately explained information. Stakeholders should receive accurate and verifiable information pertaining to environmental performance and be given a meaningful opportunity to respond to the information and make constructive suggestions for improvements.

The management representative and other individuals involved in the development and operation of the environmental management system need to have training on collaborative decision-making techniques, as explained in Chapter 12, in order to learn how to reach consensus with internal and external stakeholders. This process provides an open forum for all stakeholders to be proactively involved in reaching collaborative decisions while reducing the likelihood of internal conflicts with employees and litigation with external stakeholders.

9. Environmental Management System Documentation

ISO 14001 requires that the business establish in paper or electronic form a description of the core elements of the environmental management system and their interaction, together with directions to related documentation. Documentation of the system itself will vary according to the nature and size of the business. The documents are the most important vehicle to promote employee awareness of what is required by the system, to achieve the business's environmental targets and objectives, and to enable the evaluation of the system and environmental performance. ISO 14004 recognizes that where elements of the environmental management system are integrated with a business's overall operations, the environmental documentation should be integrated into existing documentation. The environmental management system documentation should include the environmental policy, objectives, and targets; the means by which environmental objectives and targets are achieved; a description of the key roles, responsibilities, and procedures; the location of and means to retrieve documentation; and other relevant elements of the business's environmental management system. The documentation can also discuss how the environmental management system elements are to be implemented.

The business needs to have a process set forth in the environmental management system for developing and maintaining documentation. The documentation system should be integrated with existing documentation and be available to all employees who have responsibility for maintaining records. Documentation must be managed so that the documents are readily identifiable, organized, and retained for a specified period using codes for the appropriate organization, division, function, activity, and contact person.

Documents should be periodically reviewed, revised as necessary, and approved by authorized personnel prior to issue. The current versions of relevant documents need to be available at all locations where operations essential to the effective functioning of the system are performed.

10. Document Control

A separate requirement of ISO 14001 is for procedures to be designed for the control of relevant documents for the environmental management system. The location of controlled documents must be established and known by the management representative and designated personnel. The document control system will require a periodic review, and any necessary revisions should be made with authorized approval for adequacy. When documents are created, document control procedures need to be carefully implemented so that the business can maintain its compliance record in an organized and retrievable manner and prevent destruction of relevant documents that can be used affirmatively to prevent a claim from being filed. Good document control ensures that the company demonstrates evidence of compliance and a commitment to environmental excellence. Current versions of relevant documents must be available where essential operations are performed. Obsolete documents should be promptly removed to prevent unintended use. Obsolete documents that are retained for legal or historical purposes should be suitably identified. The standard requires that documentation be legible, dated, identifiable, orderly, and retained for a specified time period. A detailed discussion of document retention management is provided in Chapter 16.

11. Operational Control

ISO 14001 requires businesses to identify those activities and operations that are associated with potential environmental impacts and that fall within the scope of the policy, objectives, and targets. These activities, including maintenance, must be carried out under controlled conditions. Businesses need to be careful not to delegate important responsibilities to individuals who are incapable of handling demanding tasks that can lead to environmental accidents or significant impacts. The NCPL Commission found that companies need to exercise due diligence to prevent the delegation of substantial discretionary authority to persons that have a propensity to engage in illegal activities. Placing an employee in a job without sufficient oversight and supervision that results in inadequate performance can lead to criminal or civil violations. In reviewing operational control issues, companies need to be particularly attuned to assigning responsible employees to sensitive positions and to give them sufficient supervision and control to achieve the goals, targets, and objectives of the environmental management system without putting the company in any unnecessary risk.

Operational control includes documenting all procedures for the handling of goods and the providing of services that have the potential to lead to significant environmental impacts. Management has to create documented procedures for any situation where the absence of procedures could lead to procedural deviations or environmental impacts. These procedures must be incorporated into the environmental management system and communicated to employees, suppliers, and contractors.

12. Emergency Preparedness and Response

ISO 14001 requires emergency plans and procedures as needed to ensure that there will be an appropriate response to unexpected or accidental incidents. DuPont, for example, supports the chemical industry's Responsible Care® program and the oil industry's Strategies for Today's Environmental Partnership which are key programs to avoid adversely impacting the environment. DuPont has committed to be prepared for emergencies and to provide leadership to assist local communities to improve their emergency preparedness. Conoco has similarly developed "emergency preparedness plans and response capabilities."

DuPont and Conoco are positive examples of how companies need to create procedures to identify the potential for accidents and emergency situations before they happen. Companies should create procedures regarding how to respond to unplanned events, and how to prevent or mitigate the environmental impacts arising from those occurrences. The procedures need to be documented, reviewed, and revised as necessary, particularly after near misses, accidents, or emergency situations. The procedures should be periodically tested where practical. Kaiser-Hill, for example, at the former Rocky Flats nuclear weapons facility that has stored plutonium, regularly runs mock raids on the facility to test its emergency preparedness to safeguard the plutonium that could be used for terrorist activities.

ISO 14004 recommends that emergency plans can document emergency organization and responsibilities; a list of key personnel; details of emergency services (e.g., fire department and spill clean-up services); internal and external communication plans; actions taken in the event of different types of emergencies; information on hazardous materials, such as material safety datasheets (including each material's potential impact on the environment and measures to be taken in the event of accidental release); and training plans and testing for effectiveness.

Consideration should be given to the EPA audit policy and the U.S. Department of Justice's *Disclosure Guidelines* as whether to self-disclose. Businesses with a strong environmental management system, and that make a timely, complete and voluntary disclosure, have the best opportunity to secure a declination or a civil, rather than criminal, penalty. Before disclosing to the government, a business should correct the violation as soon as possible or create a corrective plan; determine the legal duty to disclose; get an objective opinion of knowledgeable environmental professionals to discuss how to take immediate corrective actions and whether to disclose; consider all related and unrelated collateral consequences that may occur as a result of disclosure; and consider how to eliminate the cause of the violation by implementing an environmental system or changing the existing one to prevent recurrence of violations.

Companies that have crisis management plans, like DuPont, Conoco, and Kaiser-Hill, have taken preventive steps to avoid accidents and those types of catastrophic events that have led to strong public criticism of particular industries in the last several decades. Preventive measures can include regular maintenance and review and redesign of products, processes, and services to eliminate risk-creating circumstances. Professional risk managers and consultants can review operations and make suggestions how to avoid activities that could be construed as possibly creating opportunities for employees to engage in environmental crime.

13. Monitoring and Measuring

ISO 14001 requires a business to monitor and measure, on a regular basis, key operations, characteristics, and activities that may have significant environmental impacts. Like most of the other ISO 14001 requirements, procedures need to be documented to satisfy the standard. The business must record information to track performance, conformance to objectives and targets, and relevant operational controls. Monitoring equipment must be calibrated and maintained, with records retained for a specified period of time. There also needs to be a periodic evaluation of compliance with relevant environmental legislation and regulations.

Self-monitoring is an important concept that was studied carefully by the NCPL Commission. Specially trained compliance personnel can employ self-monitoring techniques for high-risk operations that are either subject to significant regulations or where the nature and history of such operations or facilities has suggested a significant potential for an industrial accident (NCPL, 1996). Self-monitoring may also serve as a training mechanism, as well as a check and deterrent in those situations where reportable performance measures bear on compliance levels (NCPL, 1996). Self-monitoring can be used as an early warning system to alert management of certain types of environmental problems that require changes in procedures, operations, or processes.

For small companies, self-monitoring may simply mean regular attention by line managers to environmental compliance issues on a day-to-day basis which includes managerial oversight of subordinates (NCPL, 1996). Larger companies can afford more sophisticated systems, which include monitoring mechanisms built into operational controls. Companies may wish to consider real-time monitoring for particularly important environmental aspects of operation-like emissions or toxic waste handling procedures. Real-time monitoring may mitigate harmful effects of past noncompliance and prevent repetition of the same mistakes (NCPL, 1996). Whatever the level of financial commitment, self-monitoring should be a nonintrusive method of managing risks and reducing potential liabilities.

14. Nonconformance and Corrective and Preventive Action

ISO 14001 requires businesses to define responsibility and authority to respond to and investigate nonconformance and to take action to mitigate the impacts of nonconformance. The business must initiate and complete corrective and preventive actions to eliminate the causes of actual or potential nonconformance. The standard does not state how this should be done nor in fact does ISO 14004 provide any guidance, leaving all decisions to be made in this controversial and highly charged area to the discretion of management. The standard simply specifies that the actions must be commensurate with the environmental impacts. These actions should result in the implementation and recording of procedural changes in the environmental management system resulting from the actions.

Preventive actions need to be taken in the first instance to avoid noncompliance. Participation in the EPA's voluntary programs, civil and criminal litigation awareness programs, and self-monitoring can remove many of the causes of noncompliance in the course of the development of a sound environmental management system. If an incident occurs despite all these efforts, steps must be taken immediately to minimize risks to the company.

The NCPL Commission found that companies must be proactive in this approach to dealing with incidents of noncompliance before the incident occurs. For example, a company should designate a specific individual with the responsibilities of identifying and responding to actual or suspected violations. That individual should be designated in advance of occasions for such investigations and have special procedures for gathering evidence of misconduct, as well as adequate resources available to conduct an investigation. The company should ensure the independence of the investigators from line managers whose activities or supervision may be subject to investigation. Investigators need to take steps to assure the accuracy and reliability of information gathered during the investigation and conduct interviews in a manner that is likely to preserve the attorney-client and work product privileges.

If evidence of an environmental violation is found, the company needs to make a determination whether to disclose the potential violation to authorities. The EPA and many state regulatory authorities have strict procedures including time limitations for making such disclosures to authorities. Those procedures are set forth in Chapter 11. Hence, it is important to designate decision-making responsibility and authority for determining when and how decisions to self-report detected misconduct should be made, giving sufficient latitude to the decision makers to take each incident on a case-by-case basis to avoid circumstances where a company's procedures require an action that is not warranted by the facts. Persons responsible for deciding whether to disclose have only a brief period of time, perhaps as few as ten days to make the decision. During this time, evidence has to be collected, people interviewed, and legal decisions made regarding whether a violation has actually occurred and whether the advantages of reporting the misconduct outweigh the risks of keeping that information confidential.

In analyzing the difficult and delicate factual and legal considerations with a wide range of possible negative consequences, it is also necessary to determine the appropriate scope of disclosure if a decision is made to disclose; address potential conflicts of interest between the business and its employees who may be subject to charges and who are outside the protection of government disclosure policies; and determine how to preserve legal privileges in the course of making the disclosure to authorities. It may also be necessary to decide whether to cooperate with external investigators and to determine when and how to remediate the environmental harm. Other legal considerations include identifying the scope and ramifications of the business's vicarious responsibility to third parties for the detected misconduct. With all of these decisions occurring within a brief period of time, it is important for the business to lay out the procedures in advance in order to minimize confusion, uncertainty, and inconsistent decisions if an incident arises that requires a decision whether to disclose an environmental violation to authorities.

Reliable records of the investigation and the remediation decision must be preserved to document the nonconformance, as well as the corrective and preventive action taken. In this regard, ISO 14004 recommends that the findings, conclusions, and recommendations reached as a result of measuring, monitoring, auditing, and other reviews of the environmental management system should be documented and the necessary corrective and preventive actions identified. Documenting the noncompliance is included so that evidence may be preserved for future use.

15. Records

ISO 14001 requires procedures to be documented to identify, maintain, and dispose of environmental records. This requirement needs to be carefully distinguished from environmental management system documentation, which involves maintaining in paper or electronic form the records that document the existence of that system. The ISO 14001 records requirement applies to all environmental records. It provides that records must be legible, identifiable, and traceable to the activity, product, or service of the business that involves environmental aspects and impacts. The records must be readily retrievable; protected from damage, deterioration, and loss; retained as specified; and demonstrate conformance to the standard. A detailed discussion of how to create and maintain an effective document retention system is provided in Chapter 16.

Generally, environmental records will include the company's permits, training records, monitoring data, material safety data sheets and product information, information pertaining to suppliers and contractors, environmental audits, and management reviews. ISO 14004 recommends that records of environmental aspects and their associated impacts and legislative and regulatory requirements also be included, together with inspection, calibration, and maintenance activities, as well as details of nonconformance, including incidents, complaints, and follow-up action.

Records management requires developing a means of identification, collection, indexing, filing, storage, maintenance, retrieval, retention, and disposition of all pertinent environmental records, including environmental management system documentation. Careful attention has to be paid to maintaining the integrity and confidentiality of these records in this process.

16. Environmental Management System Audit

An environmental management system needs to be audited on a periodic basis to ensure that it is being implemented effectively. The standard requires that the business develop procedures for audit scope, frequency, methodologies, responsibilities, and requirements. This program must include periodic audits to determine if the environmental management system conforms to planned arrangements, including the requirements of the standard. The audit must determine if the environmental management system is properly implemented and maintained. The audit program, including the schedule, must be based on environmental impacts and past results. Results of the environmental management system audit must be provided to management. A continual improvement process should be applied to an environmental management system to achieve overall improvement in environmental performance.

The NCPL Commission found that auditing is one of the reasonable steps for a business to take to achieve compliance with its self-imposed standards as well as requirements of law. Evaluative auditing programs can determine the effectiveness of the environmental management system. Audits can be designed to perform different purposes, such as an unannounced audit to deter and detect willful misconduct or preannounced audits which are likely to be less disruptive and detect unintentional mistakes. Auditing can be done in conjunction with self-monitoring and regular reporting functions of the company. Auditing may uncover a violation that may result in the company's need to decide how to correct the problem and whether to disclose the violation to the government to avoid penalties. The NCPL Commission provides as an example that an environmental audit can check a facility's emissions and the integrity of the monitoring processes in place.

Companies need to evaluate the desirable frequency and scope of audits and assess the independence and reliability of those individuals who perform audits. An open question is whether the company should have audits conducted by persons either inside or outside the business or a particular business unit. The NCPL Commission advised that systems need to be devised to assure follow-ups to register findings which have as a component a means for employees or agents to report violations of the standards. For those employees who do report violations, a system should be in place for them to do so anonymously, without retaliation, if they so desire. The audit program's procedures must be fully documented and include a records retention component, as discussed in Chapter 16.

The following companies have publicly disclosed that they consider auditing an important part of their efforts to manage environmental risks:

- **Union Carbide's** health, safety, and environmental audit program evaluates facility compliance with internal standards and regulatory requirements. Internal auditors visit sites and classify performance. The scale ranges from M (meets requirements) down to RSI (requires substantial improvement). Audit scheduling focuses attention worldwide on locations that have greater potential risk or did not perform up to Union Carbide standards in prior audits, even though they are in legal compliance.

An important aspect of Union Carbide's system is participation by auditors from an outside consulting firm in roughly a fourth of environmental, health, and safety audits company-wide each year. The outside auditors perform follow-up analysis and provide an annual assessment of environmental, health, and safety audit performance directly to an independent subcommittee of Union Carbide's Board of Directors. Locations are required to develop action plans to correct all deficiencies identified during audits. In addition, any location receiving an RSI rating is required to explain deficiencies, action plans for correcting them, and plans for preventing their recurrence, directly to senior management.

- **Conoco** measures its performance and learns from it by systematically auditing behaviors, work processes, management systems and equipment, and promptly correcting any deficiencies.

- **IBM** conducts rigorous audits and self-assessments of its compliance with its environmental policy, measures progress of its environmental affairs performance, and reports periodically to the Board of Directors.

- **Westinghouse** began a formal auditing program in 1988 to perform internal assessments of all facilities and provide management with verification that they are in compliance with government regulations and company policies and procedures. Environmental, safety, and industrial hygiene issues are reviewed to ensure that deficiencies are identified and promptly corrected. The Westinghouse Environmental Audit Program has developed a "world-class" reputation for quality and effectiveness.

- **Newport Electric Corp.** has agreed to conduct a periodic environmental review and will work toward the timely creation of independent environmental audit procedures. It plans to review its progress in implementing these principles and in complying with all applicable laws and regulations.

- **The Southern Company's** goal is to meet or surpass all environmental laws, regulations, and permit requirements, and verifying this commitment through environmental auditing.

- **Halliburton Company** requires audits to be conducted periodically, with priority attention given to environmentally sensitive operations and facilities. Appropriate actions will be taken to address environmental concerns.

17. Management Review

Top management must review the environmental management system to ensure continued suitability, adequacy, and effectiveness. The necessary information must be collected to allow for a meaningful review. The review must be documented; consider audit results, changing circumstances, and commitment to continual improvement; and address the need for changes to policy, objectives, and environmental management system elements.

The review of the environmental management system should include: review of environmental objectives, targets, and environmental performance; findings of the environmental management system, self-monitoring and audits; an evaluation of its effectiveness; an evaluation of the suitability of the environmental policy and the need for changes in the light of legislation, expectations, requirements of interested parties, the products or activities of the business or because of advances in science and technology; lessons learned from environmental incidents; market preferences; and reporting and communication. The following companies have disclosed how they conduct management reviews:

- **Chevron** has agreed to expand "corporate environmental compliance reviews to include assessing the effectiveness of environmental, safety, fire, and health management systems, in addition to maintaining legal compliance. This should help the Company identify and modify those processes, products and practices that may involve unacceptably high risks when judged by the standards of anticipated laws and regulations."
- **StorageTek's** "executive management will regularly review safety and environmental management performance and compliance and ensure adherence to established goals and policies."
- **Lockheed Martin** "reviews company program effectiveness in implementing corporate policy, achieving and maintaining compliance with laws and regulations, stewardship of assets and reduction of EHS costs. Facilitate the transfer of EHS information, best practices and technology."

The NCPL Commission found that such management reviews, such as Chevron's, StorageTek's, or Lockheed Martin's, may wind up disciplining or possibly retraining responsible employees and identifying root causes of noncompliance including weaknesses in detection practices. External evaluators can provide management with objective evaluations of the environmental management system and provide useful insights on how the program can be strengthened. Management reviews should be conducted regularly and promptly to assure effective follow-up measures.

Many businesses that institute ISO 14001 compliant environmental management systems desire to gain the international recognition earned by successful implementation of their environmental management system. After a business has developed and implemented its ISO 14001 environmental management system, it can retain a third-party registrar to assess the system's conformance with the standard. Chapter 20 discusses how this can be accomplished.

Chapter 20

ISO 14001 CERTIFICATION AUDITS AND REGISTRATION

The ISO 14000 standards allow for the independent third-party certification of a business's environmental management systems, ensuring both credibility and integrity. Although self-declaration of ISO 14001 conformance is allowed, many businesses will seek the benefits of third-party certification.

Any company wishing to minimize environmental liability may significantly benefit by implementing an ISO 14001-compliant environmental management system. If clients and other interested parties do not require or prefer that the system be certified, then the company may be able to save, or at least delay, the expense associated with third-party certification. However, most companies that develop ISO 14001 environmental management systems eventually realize that third-party certification affords significant recognition and benefits that outweigh the expense and inconvenience of the registration process.

The 14000 series provides a single framework recognized worldwide for strategic environmental risk management. Companies that adopt the standard will be included in a group of world-class organizations, reflecting the cutting edge of environmental risk management. Formal certification will also enhance a company's public image, decrease liability exposure, provide a mechanism for increasing process efficiency, lower operating costs and contingency set-asides, and decrease potential enforcement penalties. Continual improvement of environmental management systems will also lead to improved environmental performance.

For a company's environmental management system to become certified, a third-party registrar must conduct a series of registration audits. During the registration audit, the registrar's assessment team will conduct an on-site audit to evaluate and verify that the environmental management system has been effectively implemented and conforms to the requirements of ISO 14001. When the assessment has been successfully completed, the registrar will issue the Accredited Certificate of Approval, which is valid for three years. To maintain approval, the registrar will then conduct routine surveillance visits, normally at six-month intervals.

ACCREDITATION, CERTIFICATION, AND REGISTRATION

The terms certification, accreditation, and registration are a significant source of confusion. The International Organization for Standardization's Conformity Assessment Committee defines these terms as follows:

Accreditation: A procedure by which an authoritative body gives formal recognition that a body or person is competent to carry out specific tasks

Certification: A procedure by which a third party gives written assurance that a product, process, or service conforms to specific requirements

Registration: A procedure by which a body indicates relevant characteristics of a product, process, or service, or specific information about a body or person, and then includes or registers the product, process, or service in an appropriate publicly available list

Although the terms registration and certification have slightly different meanings, they are synonymous in common usage. In the U.S. the term registration is more commonly used, whereas the international community prefers the term certification.

Registration Procedure

Individual sites or complete companies can be registered to ISO 14001. The procedure for registering a site or company is similar to that for ISO 9000 (quality) registration. A company may request a joint quotation to cover both ISO 9000 (quality) and ISO 14000 (environmental) registration. If both quality and environmental registration are sought, a joint registration process should result in significantly reduced fees since redundancies can be avoided.

To become registered, a company will initially be required to provide information regarding the size, type of business, potential environmental impacts, and registration readiness. On the basis of that information, the registrar will provide a quotation for the cost of the preregistration assessment, the assessment audit, and the subsequent surveillance visits.

Typically, a preassessment and review of the environmental manuals will be conducted at least four weeks prior to the full assessment audit. The purpose of the preassessment audit is to establish that the company's environmental management system covers all of the ISO 14001 requirements. At the completion of the preassessment audit, a confidential report will be provided that details minor and major deficiencies. The full assessment audit will then be scheduled to allow time for the company to take any necessary corrective actions.

The full assessment audit will include a detailed examination of both the documented environmental management system and indications of the success of its implementation. During the typical one-week visit, all nonconformances will be noted and the need for any corrective actions discussed. In the case of major nonconformities, corrective action must be completed and documented prior to certification. The registrar will work with the company to clarify the nonconformities. It should be noted, however, that the registrar is not a consultant, and is thus unlikely to provide guidance regarding the method of change — just the need for improvement. When the assessment has been successfully completed, and all major nonconformances resolved, the registrar will issue the

Accredited Certificate of Approval. This certificate is valid for three years, subject to the satisfactory maintenance of the management system.

To maintain approval, the registrar monitors the company's conformance to the requirements of ISO 14001 by conducting routine records and surveillance visits, typically at six-month intervals. At the completion of the three-year certification period, the registrar will conduct a reassessment. Since the majority of the management system elements will have been addressed during the six-month, revolving-focus visits, the reassessment will typically take much less time than was taken during the initial assessment. Once registered, a company can advertise its certified or registered status by displaying the registrar's logo.

Choosing a Registrar

Since the quality and reputation of registrars vary significantly, the business should choose one carefully. Once a registrar is chosen, it can be both costly and cumbersome to change within the three-year assessment cycle. In the U.S. the American National Standards Institute (ANSI) has teamed with the Registration Accreditation Board (RAB) to accredit registrars. When choosing a registrar in the U.S., the business should require that they be ANSI/RAB accredited under the National Accreditation Program (NAP).

Registrars are accredited to conduct registration assessments by industry designation (SIC Code). If the company to be assessed predominantly manufactures electrical components, a registrar accredited to assess electrical components manufacturers would be required. If the company has multinational operations that desire or may in the future desire to become registered, a registrar that is accredited in the subsidiaries' countries will also be necessary. If additional registrations become desirable in the future, such as the Eco Management and Audit Scheme (EMAS) or ISO 9000, a registrar with these capabilities should initially be chosen. Having different registrars for different management systems would be cumbersome and expensive.

Finally, since the business relationship that develops between a company and its registrar typically lasts more than three years, a registrar with both longevity and financial security is preferred. It would be a major inconvenience for a business to invest a significant amount of money and time into working with a registrar who subsequently goes out of business. The registrar must also have a long-term commitment to the assessment of environmental management systems.

CONCLUSION

The last quarter of the 20th Century has witnessed an unprecedented change in how we value the environment. At all levels of government and throughout society, people have a greater understanding and appreciation of the need to protect our ecosystem. The only outstanding question is how much we are willing to invest, individually and collectively, in time and resources to contribute to sustainable growth for the future.

Businesses must learn how to eliminate pollution, which will reduce risks and expenses and create opportunities for revenue growth. Environmental management systems may become the most expedient and effective means of transmitting the message to everyone in the organization from the CEO and the board of directors to the lowest level employee that they have an important role in any effort minimizing environmental impacts. To take responsibility for their actions, officers, directors, and employees must understand how their individual impacts affect the environment and how they can work together in reducing future degradation. Businesses must design a meaningful message so that everyone understands that their collective performances have an extraordinary cumulative effect on the environment. One employee taking responsibility and eliminating a risk can have positive consequences and ramifications for all aspects of the business and society in general.

We should not underestimate the power of collective thinking in reaching answers to the most difficult and sensitive environmental questions that pervade our society. We cannot solve these problems in a vacuum. An ISO 14000 environmental management system, along with risk management and risk reduction techniques, is the most effective means of accomplishing this objective.

Appendix A

RIO DECLARATION ON THE ENVIRONMENT AND DEVELOPMENT

1992

The United Nations Conference on Environment and Development, having met at Rio de Janeiro from 3 to 14 June 1992, reaffirming the Declaration of the United Nations Conference on the Human Environment, adopted at Stockholm on 16 June 1972, and seeking to build upon it, with the goal of establishing a new and equitable global partnership through the creation of new levels of co-operation among States, key sectors of societies and people, working towards international agreements which respect the interests of all and protect the integrity of the global environmental and developmental system, recognizing the integral and interdependent nature of the Earth, our home, proclaims that:

Principle 1
Human beings are at the center of concerns for sustainable development. They are entitled to a healthy and productive life in harmony with nature.

Principle 2
States have, in accordance with the charter of the United Nations and the principles of international law, the sovereign right to exploit their own resources pursuant to their own environmental and developmental policies, and the responsibility to ensure that activities within their jurisdiction or control do not cause damage to the environment or other States or of areas beyond the limits of national jurisdiction.

Principle 3
The right to development must be fulfilled so as to equitably meet developmental and environmental needs of present and future generations.

Principle 4
In order to achieve sustainable development, environmental protection shall constitute an integral part of the development process and cannot be considered in isolation from it.

Principle 5
All States and people shall cooperate in the essential task of eradicating

poverty as an indispensable requirement for sustainable development, in order to decrease the disparities in standards of living and better meet the needs of the majority of the people of the world.

Principle 6
The special situation and needs of developing countries, particularly the least developed and those most environmentally vulnerable, shall be given special priority. International actions in the field of environment and development should also address the interest and needs of all countries.

Principle 7
States shall cooperate in a spirit of global partnership to conserve, protect, and restore the health and integrity of the Earth's ecosystem. In view of the different contributions to global environmental degradation, States have common but differentiated responsibilities. The developed countries acknowledge the responsibility that they bear in the international pursuit of sustainable development in view of the pressures their societies place on the global environment and of the technologies and financial resources they command.

Principle 8
To achieve sustainable development and a higher quality of life for all people, States should reduce and eliminate unsustainable patterns of production and consumption and promote appropriate demographic policies.

Principle 9
States should cooperate to strengthen endogenous capacity-building for sustainable development by improving scientific understanding through exchanges of scientific and technological knowledge, and by enhancing the development, adaptation, diffusion, and transfer of technologies, including new and innovative technologies.

Principle 10
Environmental issues are best handled with the participation of all concerned citizens, at the relevant level. At the national level, each individual shall have appropriate access to information concerning the environment that is held by public authorities, including information on hazardous materials and activities in their communities, and the opportunity to participate in decision-making processes. States shall facilitate and encourage public awareness and participation by making information widely available. Effective access to judicial and administrative proceedings, including redress and remedy, shall be provided.

Principle 11
States shall enact effective environmental legislation. Environmental standards, management, objectives, and priorities should reflect the environmental and developmental context to which they apply. Standards applied by some

countries can be inappropriate and of unwarranted economic and social cost to other countries, in particular, developing countries.

Principle 12

States should cooperate to promote a supportive and open international economic system that would lead to economic growth and sustainable development in all countries, to better address the problems of environmental degradation. Trade policy measures for environmental purposes should not constitute a means of arbitrary or unjustifiable discrimination or a disguised restriction on international trade. Unilateral actions to deal with environmental challenges outside the jurisdiction of the importing country should be avoided. Environmental measures addressing transboundary or global environmental problems should, as far as possible, be based on an international consensus.

Principle 13

States shall develop national law regarding liability and compensation for the victims of pollution and other environmental damage. States shall also cooperate in an expeditious and more determined manner to develop further international law regarding liability and compensation for adverse effects of environmental damage caused by activities within their jurisdiction or control to areas beyond their jurisdiction.

Principle 14

States should effectively cooperate to discourage or prevent the relocation and transfer to other States of any activities and substances that cause severe environmental degradation or are found to be harmful to human health.

Principle 15

In order to protect the environment, the precautionary approach shall be widely applied by States according to their capabilities. Where there are threats of serious or irreversible damage, lack of full scientific certainty shall not be used as a reason for postponing cost-effective measures to prevent environmental degradation.

Principle 16

National authorities should endeavor to promote the internalization of environmental costs and the use of economic instruments, taking into account the approach that the polluter should, in principle, bear the cost of pollution, with due regard to the public interest and without distorting international trade and investment.

Principle 17

Environmental impact assessment, as a national instrument, shall be undertaken for proposed activities that are likely to have a significant adverse impact on the environment and are subject to a decision of a competent national authority.

Principle 18

States shall immediately notify other States of any natural disasters or other emergencies that are likely to produce sudden harmful effects on the environment of those States. Every effort shall be made by the international community to help States so afflicted.

Principle 19

States shall provide prior and timely notification and relevant information to potentially affected States on activities that can have a significant adverse transboundary environmental effect and shall consult with those States at an early stage and in good faith.

Principle 20

Women have a vital role in environmental management and development. Their full participation is therefore essential to achieve sustainable development.

Principle 21

The creativity, ideals, and courage of the youth of the world should be mobilized to forge a global partnership in order to achieve sustainable development and ensure a better future for all.

Principle 22

Indigenous people and their communities, and other local communities, have a vital role in environmental management and development because of their knowledge and traditional practices. States should recognize and duly support their identity, culture, and interest and enable their effective participation in the achievement of sustainable development.

Principle 23

The environment and natural resources of people under oppression, domination, and occupation shall be protected.

Principle 24

Warfare is inherently destructive of sustainable development. States shall therefore respect international law providing protection for the environment in times of armed conflict and cooperate in its further development, as necessary.

Principle 25

Peace, development, and environmental protection are interdependent and indivisible.

Principle 26

States shall resolve all their environmental disputes peacefully and by appropriate means in accordance with the Charter of the United Nations.

Principle 27

States and people shall cooperate in good faith and in a spirit of partnership in the fulfillment of the principles embodied in this Declaration and in the further development of international law in the field of sustainable development.

Appendix B

The International
Chamber of Commerce

BUSINESS CHARTER FOR
SUSTAINABLE DEVELOPMENT

PRINCIPLES FOR ENVIRONMENTAL MANAGEMENT

1. Corporate priority
To recognize environmental management as among the highest corporate priorities and as a key determinant to sustainable development; to establish policies, programs, and practices for conducting operations in an environmentally sound manner.

2. Integrated management
To integrate these policies, programs, and practices fully into each business as an essential element of management to all its functions.

3. Process of improvement
To continue to improve corporate policies, programs, and environmental performance, taking into account technical developments, scientific understanding, consumer needs, and community expectations with legal regulations as a starting point and to apply the same environmental criteria internationally.

4. Employee education
To educate, train, and motivate employees to conduct their activities in an environmentally responsible manner.

5. Prior assessment
To assess environmental impacts before starting a new activity or project and before decommissioning a facility or leaving a site.

6. Products and services
To develop and provide products or services that have no undue environmental impact and are safe in their intended use, that are efficient in their consumption of energy and natural resources, and that can be recycled, reused, or disposed of safely.

7. Customer advice

To advise, and where relevant, educate customers, distributors, and the public in the safe use, transportation, storage, and disposal of products provided; and to apply similar considerations to the provision of services.

8. Facilities and operations

To develop, design, and operate facilities and conduct activities taking into consideration the efficient use of energy and materials, the sustainable use of renewable resources, the minimization of adverse environmental impact and waste generation, and the safe and responsible disposal of residual wastes.

9. Research

To conduct or support research on the environmental impact of raw materials, products, processes, emissions, and wastes associated with the enterprise and on the means of minimizing such adverse impacts.

10. Precautionary approach

To modify the manufacture, marketing, or use of products or services or the conduct of activities consistent with scientific and technical understanding, to prevent serious or irreversible environmental degradation.

11. Contractors and suppliers

To promote the adoption of these principles by contractors acting on behalf of the enterprise, encouraging and, where appropriate, requiring improvements in their practices to make them consistent with those of the enterprise, and to encourage the wider adoption of these principles by suppliers.

12. Emergency preparedness

To develop and maintain, where significant hazards exist, emergency preparedness plans in conjunction with the emergency services, relevant authorities, and the local community recognizing potential transboundary impacts.

13. Transfer of technology

To contribute to the transfer of environmentally sound technology and management methods throughout the industrial and public sectors.

14. Contributing to the common effort

To contribute to the development of public policy and to business, governmental, and intergovernmental programs and education initiatives that will enhance environmental awareness and protection.

15. Openness to concerns

To foster openness and dialogue with employees and the public, anticipating and responding to their concerns about the potential hazards and impacts of operations, products, wastes, or services, including those of transboundary or global significance.

16. Compliance and reporting

To maintain environmental performance, to conduct regular environmental audits and assessments of compliance with company requirements, legal requirements, and these principles, and periodically to provide appropriate information to the Board of Directors, shareholders, employees, the authorities, and the public.

Appendix C

U.S. SENTENCING GUIDELINES FOR ORGANIZATIONS

UNITED STATES SENTENCING COMMISSION

An "effective program to prevent and detect violations of law" means a program that has been reasonably designed, implemented, and enforced so that it generally will be effective in preventing and detecting criminal conduct. Failure to prevent or detect the instant offense, by itself, does not mean that the program was not effective. The hallmark of an effective program to prevent and detect violations of law is that the organization exercised due diligence in seeking to prevent and detect criminal conduct by its employees and other agents. Due diligence requires at a minimum that the organization must have taken the following types of steps:

1. The organization must have established compliance standards and procedures to be followed by its employees and other agents that are reasonably capable of reducing the prospect of criminal conduct.

2. Specific individual(s) within high-level personnel of the organization must have been assigned overall responsibility to oversee compliance with such standards and procedures.

3. The organization must have used due care not to delegate substantial discretionary authority to individuals whom the organization knew, or should have known through the exercise of due diligence, had a propensity to engage in illegal activities.

4. The organization must have taken steps to communicate effectively its standards and procedures to all employees and other agents, *e.g.,* by requiring participation in training programs or by disseminating publications that explain in a practical manner what is required.

5. The organization must have taken reasonable steps to achieve compliance with its standards, *e.g.,* by utilizing monitoring and auditing systems reasonably designed to detect criminal conduct by its employees and other agents and by having in place and publicizing a reporting system whereby employees and other agents could report criminal conduct by others within the organization without fear of retribution.

6. The standards must have been consistently enforced through appropriate disciplinary mechanisms, including, as appropriate, discipline of individuals responsible for the failure to detect an offense. Adequate discipline of individuals responsible for an offense is a necessary compo-

nent of enforcement; however, the form of discipline that will be appropriate will be case specific.

7. After an offense has been detected, the organization must have taken all reasonable steps to respond appropriately to the offense and to prevent further similar offenses — including any necessary modifications to its program to prevent and detect violations of law.

The precise actions necessary for an effective program to prevent and detect violations of law will depend upon a number of factors. Among the relevant factors are:

1. **Size of the organization** — The requisite degree of formality of a program to prevent and detect violations of law will vary with the size of the organization: the larger the organization, the more formal the program typically should be. A larger organization generally should have established written policies defining the standards and procedures to be followed by its employees and other agents.

2. **Likelihood that certain offenses may occur because of the nature of its business** — If because of the nature of an organization's business there is a substantial risk that certain types of offenses may occur, management must have taken steps to prevent and detect those types of offenses. For example, if an organization handles toxic substances, it must have established standards and procedures designed to ensure that those substances are properly handled at all times. If an organization employs sales personnel who have flexibility in setting prices, it must have established standards and procedures designed to prevent and detect price-fixing. If an organization employs sales personnel who have flexibility to represent the material characteristics of a product, it must have established standards and procedures designed to prevent fraud.

3. **Prior history of the organization** — An organization's prior history may indicate types of offenses that it should have taken actions to prevent. Recurrence of misconduct similar to that which an organization has previously committed casts doubt on whether it took all reasonable steps to prevent such misconduct.

An organization's failure to incorporate and follow applicable industry practice or the standards called for by any applicable governmental regulation weighs against a finding of an effective program to prevent and detect violations of law.

U.S. Sentencing Guidelines § 8A1.2 (November 1, 1991)

Appendix D

DENVER METRO CHAMBER OF COMMERCE

ENVIRONMENTAL POLICY STATEMENT

1. Environmental Responsibility and Leadership

To take responsibility for actions that impact the environment and to lead the business community in a manner that balances the environmental, health, safety, and economic interests of the community.

2. Cooperation Within the Industry for Progress

To work closely with business leaders to preserve and protect the environment, and to share information and create heightened awareness of environmental policy, laws, and regulations.

3. Input and Advocacy

To serve as a source of information on environmental issues which affect its members and to maintain an active role in the development of environmental policy, laws, and regulations through interaction with federal, state, and local agencies.

4. Implementation of Environmental Goals

To assign a member of senior management or a member of the Board of Directors responsibility for the successful implementation of this environmental policy statement.

5. Environment Research and Business Development

To promote the metropolitan Denver area as a center of excellence for environmental research and development.

6. Communication and Cooperation

To communicate with the community about its operations and cooperate with groups committed to the responsible protection and preservation of the environment.

7. Efficient Use of Energy

To use energy efficiently; to purchase, whenever practical, updated equipment which uses less energy; and to promote the use of new technologies and methods that maximize output while minimizing energy consumption.

8. Pollution Prevention and Waste Minimization

To prevent pollution whenever possible and to keep the production of waste generated by the Chamber to a minimum by attempting to maximize recycling, conservation, and renewable resource programs, and to promote similar activities within its membership.

9. Environmental Training and Education

To train and educate all employees concerning workplace and environmental safety principles, and to serve as a resource forum from which its members can obtain information regarding environmental issues.

10. OSHA and ADA Compliance

To comply with standards set forth by the Occupation Safety and Health Administration (OSHA) and the Americans with Disabilities Act (ADA).

REFERENCES

PREFACE

Anderson, R., The Journey from There to Here: The Eco-Odyssey of the CEO, *Presented at the Second International Green Building Conference and Exposition, Big Sky, MT*, Aug. 13-15, 1995 (on file with the authors).

Anon., Environmental Spending, *The Economist*, Oct. 12, 1991.

Clark, W., Managing Plant Earth, *Scientific American*, vol. 269, 1989.

Cortese, A.D. ScD, Earth Day 1995: Partnerships for Sustainabilities, *Presented at New England Earth Day Organizing Conference, John F. Kennedy School of Government*, Nov. 5, 1994 (on file with the authors).

Hart, S.L., Beyond Greening: Strategies for a Sustainable World, *Harvard Business Review*, Jan./Feb. 1997.

Hawken, P. and McDonough, W., Seven Steps to Doing Good Business, *Inc.*, Nov. 1994.

Lavelle, M., Some Costs Benefit Companies, *National Law Journal*, Nov. 27, 1995.

Magretta, J., Growth Through Global Sustainability, *Harvard Business Review*, Jan./Feb. 1997.

Mcinerney, F. and White, S., The Total Quality Corporation, *Truman Talley Books/Plume*, 1995.

Montalbano, W.D., Verdict Near in McLibel Food Fight, *Denver Post*, June 16, 1997.

Porter, M.E. and van der Linde, C., Green and Competitive: Ending the Stalemate, *Harvard Business Review*, Sept./Oct. 1995(a).

Porter, M.E. and van der Linde, C., Toward a New Conception of the Environment - Competitiveness Relationship, *Journal of Economic Perspectives*, Fall 1995(b).

Rothenberg, E.B., Smith, C.A. and Feitshans, I.L., Caremark International, Inc. - Directors' Obligation to Assure Compliance with Governmental Regulations, *Metropolitan Corporate Counsel*, April 1997.

Schneider, K., New View Calls Environmental Policy Misguided, *New York Times*, Mar. 21, 1993(a).

Schneider, K., Unbending Regulations Incite Move to Alter Pollution Laws, *New York Times*, Nov. 29, 1993(b).

CHAPTER 1 ISO 14000 AND RISK MANAGEMENT SYSTEMS

Begley, R., Is ISO 14000 Worth It?, *Journal of Business Strategy*, vol. 17, Sept./Oct. 1996.

Bell, C.L., ISO 14001: Application of International Environmental Management Systems Standards in the United States, *Environmental Law Reporter*, Dec. 1995.

Hall, Jr., Ridgway M., ISO 14000 Environmental Management Standards: Making the Benefits Outweigh the Burdens, *Paper Presented at the 25th ABA Annual Conference on Environmental Law, Keystone, CO*, Mar. 21-24, 1996 (on file with the authors).

CHAPTER 2 OUTSIDE IMPACTS ON CORPORATE ENVIRONMENTAL RISK MANAGEMENT PERFORMANCE

Bavaria, J., Clean Up Your Environmental Act: Withholding Investments Can Influence Corporate Actions, *Newsday*, Sept. 7, 1989.

Elkington, J., The Vintage of '96, *Tomorrow*, vol. 59, 1996.

Hall, Jr., Ridgway M., ISO 14000 Environmental Management Standards: Making the Benefits Outweigh the Burdens, *Paper Presented at the 25th ABA Annual Conference on Environmental Law, Keystone, CO*, Mar. 21-24, 1996 (on file with the authors).

Hilton, P.A. and Marlin, A.T., The Role of the Nonprofit in Rating Environmental Performance, *Corporate Environmental Strategy*, Spring 1996.

Lober, D.J., Current Trends in Corporate Reporting, *Corporate Environmental Strategy*, Winter 1997.

National Center for Preventive Law, *National Center for Preventive Law Corporate Compliance Principles, 1996*. (For a complete copy of this document, please contact: NCPL, 1900 Olive Street, Denver, CO 80220, Ph. 303-871-6099.)

CHAPTER 3 NEW APPROACHES TO RISK MANAGEMENT SYSTEMS

Anon., Report Finds Air Act's Benefits may be 70 Times Higher than Its Costs, *Environment Reporter*, Nov. 15, 1996.

Judge, W.Q., Miller, A. and Fowler, D.M., What Causes Corporate Environmental Responsiveness, *Corporate Environmental Strategy*, Spring 1996.

CHAPTER 4 MANAGING RISK USING ISO 14001: A GENERAL OVERVIEW

ISO 14001, Environmental Management Systems - Specification with Guidance for Use, *ISO/ANSI/ASTM*, 1996.

ISO 14004, Environmental Management Systems - General Guidelines on Principles, Systems and Supporting Techniques, *ISO/ANSI/ASTM*, 1996.

CHAPTER 5 BENEFITS AND COSTS OF DEVELOPING ENVIRON-MENTAL RISK MANAGEMENT SYSTEMS

Armao, J.J. and Griffith, B.J., The SEC's Increasing Emphasis on Disclosing Environmental Liabilities, *Natural Resources and Environment*, vol. 11, Spring 1997.

Mcinerney, F. and White, S., The Total Quality Corporation, *Truman Talley Books/Plume*, 1995.

CHAPTER 6 STRATEGIC INFORMATION FOR RISK MANAGEMENT SYSTEMS

Bowers, D.P., Bush, F.S., Chertow, M.R. and Hausker, K., What's All This About Reinvention?, *Environmental Forum*, Mar./Apr. 1997.

Geffen, C.A., Public Expectations and Corporate Strategy, *Corporate Environmental Strategy*, Spring 1996.

Hart, S.L., Beyond Greening: Strategies for a Sustainable World, *Harvard Business Review*, Jan./ Feb. 1997.

CHAPTER 7 ENVIRONMENTAL RISK COMMUNICATION

Morris, J. and Scarlett, L., Buying Green: Consumers, Products Labels, and the Environment, *Reason Foundation*, 1996.

CHAPTER 8 INSURANCE AND RISK TRANSFER STRATEGIES

Bailey, K.D. and Gulledge, W., Using Environmental Insurance to Reduce Environmental Liability, *Natural Resources and Environment*, Spring 1997.

Reich, D., The Merit Partnership for Pollution Prevention - An Overview, *Paper Presented at the Inter-Pacific Bar Association 7th Annual Meeting and Conference, Kuala Lumpur, Malaysia*, April 30, 1997 (on file with the authors).

CHAPTER 9 MANAGING BROWNFIELDS RISKS USING ISO 14000

Abelson, N. and McCaffrey, M., Brownfields: Recent Massachusetts and Federal Developments, *Environment Reporter*, Mar. 15, 1996.

Anon., Worth Considering, *Colorado Commons*, Spring 1997.

Brownfields Forum, Final Report and Action Plan, Oct. 1995 (on file with the authors).

Charles, L., Community Involvement with Environmental Cleanup and Economic Development, *Paper Presented at the 1997 Connecticut Industrial Site Recycling Conference*, March 14, 1997 (on file with the authors).

Freeman, D.J. and Belcamino, G.R., Brownfields Redevelopment and ISO 14000: A Marriage that Makes Sense, *Corporate Environmental Strategy*, Spring 1996.

Gore, A. (Vice President), Brownfields are Common Ground, *Brownfields News*, Apr. 1997.

Kunstler, J., Home from Nowhere, *Simon & Schuster*, 1996.

Tyson, R., Empty Factory Sites: Land of Opportunity, *USA Today*, Aug. 30, 1996.

Wilson, W.J., When Work Disappears, *Alfred A. Knopf*, 1996.

CHAPTER 10 REDUCING RISKS USING ENVIRONMENTAL POLICY

Judge, W.C., Miller, A., and Fowler, D.M., What Causes Corporate Environmental Responsiveness, *Corporate Environmental Strategy*, Spring 1996.

Kennedy, R., The Commitment to Environmental Excellence, *Presented at Corporate Stewardship of the Environment,* Report No. 982, Conference Board, 1991.

Mulligan, W., Corporate Environmentalism: Getting the Word Out, *Presented at Corporate Stewardship of the Environment* , Report No. 982, Conference Board, 1991.

CHAPTER 11 REDUCING RISKS BY AUDITING AND DISCLOSING VIOLATIONS

Cushman, J.H., Virginia Seen as Undercutting U.S. Environmental Rules, *New York Times*, Jan. 19, 1997.

EPA, Enforcement and Compliance Assurance Accomplishments Report, FY 1995, Mar 12, 1996.

EPA, EPA Audit Policy Update, Jan. 1997.

CHAPTER 12 REDUCING LITIGATION RISKS AND COSTS

Anderson, Devries, and Rodriguez, A Policyholder's Primer on Environmental Insurance Recovery, *Journal of Environmental Law and Practice*, vol. 5, 1994.

Coldiron, M.D. and Bryan, C.M., Use of Experts in Environmental Litigation and Enforcement Matters, *Natural Resources and Environment*, Summer 1996.

Haig, R.L., Corporate Counsel's Guide, Second Edition, *New York State Bar Association*, 1997.

Helmstetter, C.H., Environmental Litigation Against the Federal Government, *Natural Resources and Environment*, Summer 1996.

Mininberg, M. and Goodbody, K.S., The Superfund Rises in the Federal Courts: A Case Study, *Environment Reporter*, May 20, 1994.

CHAPTER 13 SOLVING ENVIRONMENTAL PROBLEMS USING ALTERNATIVE DISPUTE RESOLUTION TECHNIQUES

Carver, T.B. and Vondra, A.A., Alternative Dispute Resolution: Why it Doesn't Work and Why it Does, *Harvard Business Review*, May/June 1994.

Dauer, E.A., Manual of Dispute Resolution, *McGraw-Hill, Inc.*, 1994.

Dean, M., Contaminated Sites: Connecticut Moves Toward Private ADR, *Dispute Resolution Journal*, Jan./Feb. 1996.

EPA, Superfund Enforcement Mediation, Regional Pilot Project Results, 22E-2201 Oct. 1991.

Schotland, S.D., Mediation and Arbitration of Toxic Product Liability Cases, *BNA Analysis and Perspective*, Aug. 9, 1995.

Schiffer, L.J. and Juni, R.L., Alternative Dispute Resolution in the U.S. Department of Justice, *Natural Resources and Environment*, Summer 1996.

CHAPTER 14 VOLUNTARY PROGRAMS AND OTHER INITIATIVES TO REDUCE ENVIRONMENTAL RISK

EPA, Risk Reduction Through Voluntary Programs, Office of Inspector General (Report of Audit), *EIKAF6-05-0080-7100130*, Mar. 19, 1997(a).

EPA, Project XL, *EPA 231-F-97-003*, April 1997(b).

GEMI, Industry Incentives for Enviromental Improvement: Evaluation of U.S. Federal Initiatives, Sept. 1996.

Reich, D., The Merit Partnership for Pollution Prevention - An Overview, *Paper Presented at the Inter-Pacific Bar Association 7th Annual Meeting and Conference, Kuala Lumpur, Malaysia*, April 30, 1997 (on file with the authors).

CHAPTER 15 AVOIDING THE RISK OF COMMITTING AN ENVIRONMENTAL CRIME

Devaney, E.E., The Exercise of Investigative Discretion, *Memo from the EPA's Office of Criminal Enforcement*, Jan. 12, 1994.

Gaynor, K.A., Kamenar, P.D., Muchnicki, E.D., and Thomson, P., Doing Time for Environmental Crimes, *Environmental Forum*, June 23, 1993.

Guerci, L.S. and Hemphill, M., Report of Advisory Work Group on Sentencing Guidelines for Organizations Convicted of Environmental Crimes, Dissenting Views; Dec. 8, 1993 (on file with the authors).

Levin, M.E., Prosecution of Environmental Crimes in Massachusetts, *Massachusetts Department of Attorney General*, 1991.

Muchnicki, E.D., Only Criminal Sanctions Can Ensure Public Safety, *Environmental Forum*, June 23, 1993.

Ogren, R.W., The Department of Defense Voluntary Disclosure Program, *White Collar Crime*, 1991.

Pitt, H. and Groskaufmanis, K., Minimizing Corporate Civil and Criminal Liablity: A Second Look at Corporate Codes of Conduct, *Georgetown Law Journal*, vol. 78, 1990.

Thornburgh, R., Our Blue Planet: A Law Enforcement Challenge, *Keynote Address by Dick Thornburgh, Attorney General of the U.S. to the 1991 Environmental Law Enforcement Conference, New Orleans*, Jan. 8, 1991 (on file with the authors).

U.S. Department of Justice, Bureau of Justice Statistics, 1984.

Weisenbeck, T.L. and Casavechia, R.M., Guidelines for Prosecution of Environmental Violations: The Tension Between Self-Reporting and Self-Auditing, *Environment Reporter*, Mar. 6, 1992.

CHAPTER 16 USING DOCUMENT CONTROL SYSTEMS TO REDUCE RISK

Kaplan, J., Murphy, J., and Swenson, W., Compliance Programs and the Corporate Sentencing Guidelines, *Clark Boardman Callaghan*, 1994.

Skupsky, Donald E., Records Retention Procedures, *Information Requirements Clearinghouse*, 1994.

CHAPTER 17 ENVIRONMENTAL PERFORMANCE INNOVATIONS THAT REDUCE RISKS AND LIABILITIES

Anderson, R., The Journey from There to Here: The Eco-Odyssey of the CEO, *Presented at the Second International Green Building Conference and Exposition, Big Sky, MT*, Aug. 13-15, 1995 (on file with the authors).

Cortese, A.D. ScD, Earth Day 1995: Partnerships for Sustainabilities, *Presented at New England Earth Day Organizing Conference, John F. Kennedy School of Government*, Nov. 5, 1994 (on file with the authors).

Frankel, C., The Visions Gap, *Tommorrow*, July/Sept. 1995.

Hart, S.L., Beyond Greening: Strategies for a Sustainable World, *Harvard Business Review*, Jan./Feb. 1997.

Magretta, J., Growth Through Global Sustainability, *Harvard Business Review*, Jan./Feb. 1997.

Mcinerney, F. and White, S., The Total Quality Corporation, *Truman Talley Books/Plume*, 1995.

Porter, M.E. and van der Linde, C., Green and Competitive: Ending the Stalemate, *Harvard Business Review*, Sept./Oct. 1995(a).

Porter, M.E. and van der Linde, C., Toward a New Conception of the Environment - Competitiveness Relationship, *Journal of Economic Perspectives*, Fall 1995(b).

CHAPTER 18 GAPS ANALYSIS

ISO 14004, Environmental Mangement Systems - General Guidelines on Principles, Systems and Supporting Techniques, *ISO/ANSI/ASTM*, 1996.

Media Group, The ISO 14000 in Focus: A Business Perspective for Sound Environmental Management, *Video Training Workbook, CEEM*, 1995.

CHAPTER 19 IMPLEMENTING THE SYSTEMS APPROACH

Mcinerney, F. and White, S., The Total Quality Corporation, *Truman Talley Books/Plume*, 1995.

National Center for Preventive Law, *National Center for Preventive Law Corporate Compliance Principles*, 1996. (For a complete copy of this document, please contact: NCPL, 1900 Olive Street, Denver, CO 80220, Ph. 303-871-6099.)

Spedding, L.S., Environmental Management for Business, *John Wiley & Sons*, 1996.

CHAPTER 20 ISO 14001 CERTIFICATION AUDITS AND REGISTRATION

ISO 14004, Environmental Management Systems - General Guidelines on Principles, Systems and Supporting Techniques, *ISO/ANSI/ASTM*, 1996.

Media Group, The ISO 14000 in Focus: A Business Perspective for Sound Environmental Management, *Video Training, Workbook, CEEM*, 1995.

INDEX

W

waivers of subrogation (*See* insurance, environmental)

waste/s

disposal of 54, 101, 134, 173, 182, 206, 208

exchanges 190

hazardous 150, 153, 169, 190

management 32

minimization of 103

See recycling

reduction or elimination of 13, 62, 156, 187, 188, 189, 191, 205, 206, 208, 211

removal of 113

water use, reduction of 152, 153

Western Colorado Congress 122

Westinghouse 19, 100, 104, 211, 226

Westvaco 19

Weyerheuser Company 153

WG (working group) (*See* ISO, development of standards)

whistleblower 119

WMX Technologies, Inc. 99, 102, 106, 109, 110, 211

Woodall, Mark 117

working group (WG) (*See* ISO, development of standards)

World Wide Web 63, 71

W.R. Grace & Co. 163

X

Xerox 100, 102, 103, 191, 202, 203, 206, 213

XL (*See* Project XL)